Swarm Intelligence
From Social Bacteria to Humans

Editor
Andrew Schumann
University of Information Technology and Management
in Rzeszow, Poland

CRC Press
Taylor & Francis Group
Boca Raton   London   New York

CRC Press is an imprint of the
Taylor & Francis Group, an **informa** business

A SCIENCE PUBLISHERS BOOK

CRC Press
Taylor & Francis Group
6000 Broken Sound Parkway NW, Suite 300
Boca Raton, FL 33487-2742

Version Date: 20200319

International Standard Book Number-13: 978-0-367-13793-9 (Hardback)

Visit the Taylor & Francis Web site at
http://www.taylorandfrancis.com

and the CRC Press Web site at
http://www.routledge.com

Preface

Recently, in computer science there are developed different multi-agent systems which are inspired by intelligent behaviors of swarms: ants, bees, flocking of sheep (horses), shoaling and schooling of fish, etc. For example, these systems can be represented as groups of robots working together on the same task. So, swarms are regarded as a natural kind of multi-agent systems whose members support each other and have a division into different social roles in realizing their joint work. Therefore, swarms give a natural bio-inspired model of artificial multi-agent systems with some social functions.

The aim of designing groups of agents as swarms is promising today in the context of developing systems of Internet of Things, where artificial objects such as computing devices, mechanical or digital machines possess unique identifiers and can transfer data over a network without requiring an interaction with humans. Swarm intelligence can provide the Internet of Things with a new technology of modeling some social functions of these things.

For instance, we may assume that the Internet of Things can possess its own social system, where a group of objects can behave as a real swarm. The point is that swarms effectively solve different logistic tasks. So, ants and bees can solve the travelling salesman problem and bees can solve the assignment problem. Hence, we can assume that the efficiency in a self-organized realization of logistic tasks in the Internet of Things can give a benefit in the future in designing new multi-agent systems. Hence, swarm intelligence is an important branch of computer science helping us in developing artificial multi-agent systems with some social functions.

This book contains results of nine researches about swarm intelligence with an emphasis on modeling different reactions of swarms and some ways how these models of reactions can be used in computer science.

Andrew Schumann

Contents

Introduction

Andrew Schumann

Department of Cognitive Science and Mathematical Modelling,
University of Information Technology and Management in Rzeszow,
Sucharskiego 2, 35-225 Rzeszow, Poland
Email: andrew.schumann@gmail.com

The notion of swarm intelligence [6, 41] was introduced to describe the decentralized and self-organized behaviors of groups of animals. This idea was then extrapolated to design groups of robots which interacted locally to cumulate a collective reaction. Some natural examples of swarms are as follows [38]: ant colonies, bee colonies, fish schooling, bird flocking, horse herding, bacterial colonies, multinucleated giant amoebae *Physarum polycephalum*, etc. In all these examples, individual agents behave locally with an emergence of their common effect.

At first, swarm intelligence was studied in order to develop new algorithms in transporting and scheduling – the point being that ants, bees, some social bacteria, *Physarum polycephalum*, etc. can solve logistic problems very effectively [38]: (i) the Travelling Salesman Problem can be solved by ants and by amoebae; (ii) the Steiner Tree Problem can be solved by amoebae; (iii) the Generalized Assignment Problem can be solved by bees; (iv) mazes can be solved by ants and by amoebae, etc. intelligent behavior of swarm individuals is explained by the following biological reactions to attractants and repellents [5, 14, 30, 35]. Attractants are biologically active things, such as food pieces or sex pheromones, which attract individuals of the swarm. Repellents are biologically active things, such as predators, which repel individuals of the swarm. As a consequence, attractants and repellents stimulate the directed movement of swarms towards and away from the stimulus, respectively.

It is worth noting that a group of people, such as pedestrians, follow some swarm patterns such as flocking or schooling. For instance, humans prefer to avoid a person considered by them as a possible predator and if a substantial part of the group in the situation of escape panic (not less than 5%) it changes the direction, then the rest of the group follows the new direction, too. Some swarm patterns are observed among human beings under the conditions addictive behavior such as the behavior of alcoholics or gamers [38].

The methodological framework of studying swarm intelligence is represented by unconventional computing, robotics, and cognitive science. In this book we aim to analyze new methodologies involved in studying swarm intelligence. We are going to bring together computer scientists and cognitive scientists dealing with swarm patterns from social bacteria to human beings.

In modeling swarms, we assume that animal collectives can contain different numbers of their members – from a small number to a large enough number. For example, one cluster of naked mole-rats includes on average from 75 to 80 individuals, while there are ant colonies consisting of many million worker ants and many thousand queen ants living in many thousand nests. The task of simulating multi-agent systems with many millions of actors evidently is quite hard. But standard networks, we deal with in our life, such as social networks contain so many active individuals, too.

A neuromorphic computer, to which we have devoted the first contribution to this book entitled *Swarm Intelligence for Morphogenetic Engineering*, can be represented as a swarm with a huge number of active components. For designing this computer we should set 100 billion neurons and define 100 trillion nonrandom connections among them. These neurons are regarded by Bruce J. MacLennan and Allen C. McBride (the two authors of the chapter) as separate microscopic agents (microrobots) that can emit and respond to simple signals and implement simple control processes, but they can also move and transport other components. Signaling molecules and structural components are considered passive components, because they cannot move without external forces. Microrobots as active components take part in an artificial morphogenesis through assembling passive components into a desired structure. This morphogenesis is a result of interactions of collectives of microrobots.

In this chapter, Bruce J. MacLennan and Allen C. McBride proposed certain algorithms for the coordination of microrobot swarms involved in morphogenetic engineering. The morphogenetic programming notation for this purpose is based on a mathematical notation developed for partial differential equations. For more details on neuromorphic computers with artificial morphogenesis, please see [24, 25, 26, 27, 28, 29]. Hence, the ideas of Bruce J. MacLennan and Allen C. McBride appeals to the modeling and controlling of artificial swarms consisting of millions microrobots.

Some computational tasks which are being solved by swarms effectively such as transporting and scheduling are studied in depth. Nevertheless, there are many sophisticated tasks solved by swarms daily which are little known in computer science, yet. For instance, among social insects we can observe necrophoresis – a social phenomenon of carrying dead bodies of members of colonies of ants or bees from the nest [8, 15]. In the second contribution to the book under the title *Ant Cemeteries as a Cluster or as an Aggregate Pile* prepared by Tomoko Sakiyama, there is an examination of a formation of cemeteries performed by ant workers. This social behavior can be formalized by some simple clustering rules such as the following implication: if ant workers find a corpse, then they pick up it with a probability that decreases according to the cluster size, while corpse-carrying ants drop their carrying corpses with a probability that increases due to the cluster size. On the basis of these rules, ant workers build large piles of corpses [37, 40]. In the model proposed by Tomoko Sakiyama, agents can modify the probability of the drop, which was dependent on whether they detected or did not detect their nest-mates. Therefore, this chapter shows that swarms of ants comply with many objectives simultaneously from looking for food to building cemeteries.

In a flock of birds and school of fish, individuals try to coordinate their behavior on the basis of their neighbors to avoid collisions with them. However, there are many examples of group behavior without this mechanism. The soldier crabs of *Mictyris guinotae* behave as a swarm with an internal noise and/or anticipation. In the chapter *Robust Swarm of Soldier Crabs, Mictyris guinotae, Based on Mutual Anticipation* which was prepared by Y.-P. Gunji, H. Murakami, T. Niizato, Y. Nishiyama, K. Enomoto, A. Adamatzky, M. Toda, T. Moriyama and T. Kawai, there is a kinetic model analyzing how crabs move, revealing dynamic internal structures within groups, such as topological distances, scale-free correlations and inherent noise. To learn more about this model, see [12, 13].

In computer science, there are in general some basic patterns formalized of different swarms: Ant Colony Optimization [9, 10, 11], Artificial Bee Colony [16, 17, 18], Particle Swarm Optimization [19, 20], etc. The algorithms of Ant Colony Optimization and Artificial Bee Colony are aimed, first of all, for solving logistic problems such as scheduling, assignment, and transport. The algorithms of Particle Swarm Optimization allows us to simulate the group movement of animals with collision avoidance (individuals avoid a collision with neighbors), velocity matching (individuals synchronize their speed with their neighbors), and swarm centering (individuals stay close to their neighbors). All these algorithms formalizing the fundamental swarm patterns of animals can be applied in different areas of computer science: from designing artificial swarms of robots to cybersecurity. In the contribution to our book entitled *Swarm intelligence in Cybersecurity* submitted by Cong Truong Thanh, Quoc Bao Diep, and Ivan Zelinka, there are analyzed prospects of applying swarm intelligence techniques such as Ant Colony Optimization, Particle Swarm Optimization, and S-Cuckoo Search in combating cyber-attacks [2, 3, 21].

They have also considered some of the limitations of existing approaches as well as the scope for future work.

Swarm algorithms such as Ant Colony Optimization, Artificial Bee Colony, and Particle Swarm Optimization demonstrate that the main features of all swarms consist in some stable patterns of many active elements. These patterns can be well formalized within cellular automata. In the submission *Emergence of Complex Phenomena in a Simple Reversible Cellular Space* written by Kenichi Morita, we are introduced to an elementary triangular partitioned cellular automaton (ETPCA) to design reversibility [31]. In this automaton, each of its local functions is described by only four local transition rules [32]. All examples of stable patterns are provided in this chapter within a specific reversible ETPCA 0347, where 0347 is the identification number in the class of ETPCAs. On the basis of these patterns defined in ETPCA 0347, Kenichi Morita proposed a reversible logic and reversible Turing machines realized in the cellular space.

Patterns which are studied in this chapter are divided into the following three categories: periodic patterns, space-moving patterns (spaceships), and expanding patterns [32]. Some basic examples of periodic patterns are as follows: block, fin, rotator, and glider. In space-moving patterns, after some time steps p ($p > 0$) the same pattern appears at a different position. In expanding patterns, the diameter of the pattern grows indefinitely as it evolves. The patterns of all three categories can be used in designing reversible computers (reversible Turing machines) [33]. Thus, the swarm patterns can be engaged in designing bio-inspired computation techniques.

Each swarm can be represented as a network of active elements connected with each other. With this meaning, social networks can be considered a human example of swarms consisting of a set of actors and a set of dyadic connections between them. In the chapter under the title *Rough Sets over Social Networks* contributed by Krzysztof Pancerz and Piotr Grochowalski, there is an examination of a social structure in a scientific collaboration implemented in the form of papers published jointly. Such a collaboration is described by the so-called author collaboration graph. For each node of this graph there is defined its inter-team neighbourhood. Then Krzysztof Pancerz and Piotr Grochowalski appeals to rough set theory [34, 36, 39] for assessing the cohesion of co-authorships between research teams. So, they define the lower and upper inter-team neighbourhood approximations to introduce the cohesion measure of co-authorships between research teams. So, rough sets allow us to define certain properties of teams after analyzing stable patterns.

The fact that swarms prefer to follow certain stable patterns can be used in constructing bio-inspired unconventional computers on swarms: computers on ant colonies [7], computers on *Physarum polycephalum* (a multinucleated plasmodium) [1, 38, 39], etc. In the chapter *Logical*

Functions as an Idealization of Swarm Basic Reactions prepared by myself the Editor, it shows that negation, conjunction, and disjunction as basic logical functions can be regarded as some stable patterns of swarms, too. So, the logical negation can be represented as an effect of repellents on the behavior of swarm individuals, the logical conjunction as an effect of lateral inhibition (the situation of stress) because of some stimuli, and the logical disjunction as lateral activation (the situation of safety) because of some stimuli. Hence, logic is not our theoretical glasses in analyzing swarms, but it is implemented by swarms themselves.

Recently, in robotics there are actively used artificial swarms of robots to program a collective behavior emerged from the interactions between singular robots. And there appear many problems of controlling communications and interactions among these artificial agents. In the chapter *On the Motion of Agents with Directional Antennae* submitted by Alexander Kuznetsov, possible methods are examined with regard to organizing a control of connectivity for agents such as an unmanned aerial vehicle [4]. The problem is that if agents have directional antennae, it means that they have significant restrictions on the exchange of information about their location and the maintenance of a telecommunication network [22, 23]. In the process of their movement, robots must continuously change the direction of their antennae to ensure continuous communication. In this manner, the agents send messages to their nearest neighbors about the need to change locations and further such information is distributed along a chain of neighboring agents.

In the last submission under the title *Induction and Physical Theory Formation as well as Universal Computation by Machine Learning* written by Alexander Svozil and Karl Svozil, a hypothesis is put forward that in a formation of physical theories machine learning can be applied as a general, systematic framework for a generation of formal models involved in physical description and prediction. And for this machine learning some swarm algorithms can be used within deep forward networks. In other words, in future physical theories can be thought up by artificial swarms.

As we see, swarm intelligence is a large branch of studies in computer science which is united by an interest in exploring some stable patterns of natural multi-agent systems. This book contains submissions with some fresh eyes on the subject and I hope that it will be interesting for general readers of computer science.

References

1. A. Adamatzky, V. Erokhin, M. Grube, T. Schubert and A. Schumann. Physarum chip project: growing computers from slime mould. International Journal of Unconventional Computing 8(4), 319–323 (2012).

2.   Z. Ali and T.R. Soomro. An efficient mining based approach using PSO selection technique for analysis and detection of obfuscated malware. Journal of Information Assurance & Cyber Security 2018, 1–13 (2018).

3.   R. Aswani, A.K. Kar and P.V. Ilavarasan. Detection of spammers in twitter marketing: a hybrid approach using social media analytics and bio inspired computing. Information Systems Frontiers 20(3), 515–530 (2018).

4.   Scott G. Bauer, Matthew O. Anderson and James R. Hanneman. Unmanned aerial vehicle (UAV) dynamic-tracking directional wireless antennas for low powered applications that require reliable extended range operations in time critical scenarios. Technical Report INL/EXT-05-00883, Idaho National Laboratory, (October 2005).

5.   E. Ben-Jacob. Social behavior of bacteria: from physics to complex organization. Eur. Phys. J. B 65(3), 315–322 (2008).

6.   E. Bonabeau, M. Dorigo and G. Theraulaz. Swarm Intelligence: From Natural to Artificial Systems. Oxford University Press (1999).

7.   Carlos A. Coello Coello, Rosa Laura G. Zavala, Benito Mendoza García and Arturo Hernández Aguirre. Ant colony system for the design of combinational logic circuits. *In*: J. Miller, A. Thompson, P. Thomson, T.C. Fogarty (eds), Evolvable Systems: From Biology to Hardware. ICES 2000. Lecture Notes in Computer Science, vol. 1801, pp. 21–30. Springer, Berlin, Heidelberg (2000).

8.   L. Diez, P. Lejeune and C. Detrain. Keep the nest clean: survival advantages of corpse removal in ants. Biology Letters 10(7), 20140306–20140306 (2014).

9.   M. Dorigo and T. Stutzle. Ant Colony Optimization. MIT Press (2004).

10. M. Dorigo, E. Bonabeau and G. Theraulaz. Ant algorithms and stigmergy. Future Generation Computer Systems 16(8), 851–871 (2000).

11. M. Dorigo, V. Maniezzo and A. Colorni. Positive feedback as a search strategy (1991).

12. Y.-P. Gunji, Y. Nishiyama and A. Adamatzky. Robust soldier crab ball gate. Complex Systems 20, 94–104 (2011).

13. Y.P. Gunji, H. Murakami, T. Tomaru and V. Vasios. Inverse Bayesian inference in swarming behavior of soldier crabs. Philosophical Transaction of the Royal Society A 376, 20170370 (2018).

14. G.R. Ivanitsky, A.S. Kunisky and M.A. Tzyganov. Study of 'target patterns' in a phage bacterium system. *In*: Krinsky, V. (ed.), Self-organization: Autowaves and Structures Far From Equilibrium, pp. 214–217. Springer, Heidelberg (1984).

15. C. Jost, J. Verret, E. Casellas, J. Gautrais, M. Challet, J. Lluc, S. Blanco, M.J. Clifton and G. Theraulaz. The interplay between a self-organized process and an environmental template: corpse clustering under the influence of air currents in ants. Journal of The Royal Society Interface 4(12), 107–116 (2007).

16. D. Karaboga. An idea based on honey bee swarm for numerical optimization. Technical report-tr06, Engineering Faculty, Computer Engineering Department, Erciyes University (2005).

17. D. Karaboga and B. Akay. A comparative study of artificial bee colony algorithm. Appl. Math. Comput. 214(1), 108–132 (2009).

18. D. Karaboga and B. Basturk. A powerful and efficient algorithm for numerical function optimization: artificial bee colony (ABC) algorithm. Journal of global optimization 39(3), 459–471 (2007).

19. J. Kennedy and R. Eberhart. Swarm Intelligence. Morgan Kaufmann Publishers, Inc. (2001).

20. J. Kennedy and R. Eberhart. Particle swarm optimization (PSO). *In*: Proc. IEEE International Conference on Neural Networks, pp. 1942–1948. Perth, Australia (1995).

21. T. Kumaresan and C. Palanisamy. E-mail spam classification using s-cuckoo search and support vector machine. International Journal of Bio-Inspired Computation 9(3), 142–156 (2017).

22. A.V. Kuznetsov. Organization of the system of agents with the help of cellular automaton. Control of large systems 70, 136–167 (2017).

23. Alexander Kuznetsov. Self-organization of the communication network. Journal of Physics: Conference Series 1203:012093, Apr 2019.

24. Bruce J. MacLennan. Artificial Morphogenesis as an Example of Embodied Computation. International Journal of Unconventional Computing 7.1–2, 3–23 (2011).

25. Bruce J. MacLennan. Coordinating Massive Robot Swarms. International Journal of Robotics Applications and Technologies 2.2, 1–19 (2014).

26. Bruce J. MacLennan. Coordinating swarms of microscopic agents to assemble complex structures. *In*: Ying Tan (ed.), Swarm Intelligence, vol. 1: Principles, Current Algorithms and Methods. PBCE 119. Institution of Engineering and Technology, Chap. 20, pp. 583–612 (2018).

27. Bruce J. MacLennan. Embodied computation: applying the physics of computation to artificial morphogenesis. Parallel Processing Letters 22.3, p. 1240013 (2012).

28. Bruce J. MacLennan. Models and Mechanisms for Artificial Morphogenesis. *In*: F. Peper, H. Umeo, N. Matsui and T. Isokawa (eds), Natural Computing. Springer series, Proceedings in Information and Communications Technology (PICT) 2, pp. 23–33. Tokyo: Springer (2010).

29. Bruce J. MacLennan. Molecular Coordination of Hierarchical Self Assembly. Nano Communication Networks 3.2 (June), 116–128 (2012).

30. M. Margenstern. Bacteria inspired patterns grown with hyperbolic cellular automata. HPCS 757–763 (2011).

31. K. Morita. Theory of Reversible Computing. Springer, Tokyo (2017).

32. K. Morita. A universal non-conservative reversible elementary triangular partitioned cellular automaton that shows complex behavior. Natural Computing (in press). Slides with movies of computer simulation: Hiroshima University Institutional Repository, http://ir.lib.hiroshima-u.ac.jp/00039321.

33. K. Morita and R. Suyama. Compact realization of reversible Turing machines by 2-state reversible logic elements. *In*: O.H. Ibarra, L. Kari and S. Kopecki (eds), Proc. UCNC 2014, LNCS 8553, pp. 280–292 (2014.) Slides with figures of computer simulation: Hiroshima University Institutional Repository, http://ir.lib.hiroshima-u.ac.jp/00036076.

34. K. Pancerz. Quantitative assessment of ambiguities in plasmodium propagation in terms of complex networks and rough sets. *In*: T. Nakano and A. Compagnoni (eds), Proceedings of the 10th EAI International Conference on Bio-inspired Information and Communications Technologies (BICT'2017), pp. 63–66, Hoboken, USA (2017).

35. K.M. Passino. Biomimicry of bacterial foraging for distributed optimization and control. Control Syst. 22(3), 52–67 (2002).
36. Z. Pawlak. Rough Sets: Theoretical Aspects of Reasoning about Data. Kluwer Academic Publishers, Dordrecht (1991).
37. T. Sakiyama. Ant droplet dynamics evolve via individual decision-making. Scientific Reports 7, 14877-1– 14877-8 (2017).
38. A. Schumann. Behaviourism in Studying Swarms: Logical Models of Sensing and Motoring. Emergence, Complexity and Computation, vol. 33. Springer, Cham (2019).
39. A. Schumann and K. Pancerz. High-Level Models of Unconventional Computations: A Case of Plasmodium. Springer International Publishing (2019).
40. G. Theraulaz, E. Bonabeau, S.C. Nicolis, R.V. Solé, V. Fourcassié, S. Blanco, R. Fournier, J.L. Joly, P. Fernández, A. Grimal, P. Dalle and J.L. Deneubourg. Spatial patterns in ant colonies. Proc. Natl Acad. Sci. USA 99, 9645–9649 (2002).
41. I. Zelinka and G. Chen (eds). Evolutionary Algorithms, Swarm Dynamics and Complex Networks: Methodology, Perspectives and Implementation, vol. 26. Springer (2017).

Swarm Intelligence for Morphogenetic Engineering

Bruce J. MacLennan[1]* and Allen C. McBride[2]

[1] University of Tennessee, Knoxville
[2] University of Tennessee, Knoxville

1. Self-assembly of Complex Hierarchical Structures

We are applying swarm intelligence to the coordination of microrobot swarms in order to assemble complex, hierarchically structured physical systems. We are interested in multiscale systems that are intricately and specifically structured from the microscopic level up through the macroscopic level: from microns to meters. On the one hand, conventional manufacturing techniques, including additive manufacturing, work well at the macroscopic level, but are difficult to scale down below the millimeter scale. On the other hand, technologies that are effective at the microscopic scale, such as photolithography and molecular self-assembly, do not scale well to complex macroscopic objects.

Nevertheless, it would be valuable to be able to assemble automatically systems structured across many length scales. For example, the human brain has complex structures spanning more than six orders of magnitude, from micron-scale synapses to decimeter-scale functional regions and interconnections [18]. It is not unreasonable to suppose that a neuromorphic computer with capacities and functions similar to a human brain would have similar complexity. How would we assemble a neuromorphic computer with 100 billion neurons and perhaps 100 trillion nonrandom connections? Similarly complex future sensors and actuators, with capabilities similar to animal sense organs and effectors, will span scales from the microscopic to the macroscopic.

Automatic assembly of such complex structures might seem to be an unobtainable goal, but we know it is possible, as it occurs in nature. Embryos develop from a single cell into a complex organism with

*Corresponding author: maclennan@utk.edu

many trillions of cells, each itself a complex hierarchical system. During development, cells communicate, coordinate, and cooperate, behaving as a massive swarm, to differentiate and rearrange into the various interrelated tissues that constitute a complete organism. *Morphogenesis* refers to the developmental process that assembles and organizes three-dimensional forms and structures. The process is massively parallel, distributed, and robust: these are also goals for us.

Morphogenetic engineering or *artificial morphogenesis* takes inspiration from biological morphogenesis, seeking new ways to create complex structures that share some ideas from the development of embryos in multicellular organisms. Because of the diversity of mechanisms in biological morphogenesis from which inspiration might be taken, morphogenetic engineering is diverse. It can overlap several fields including swarm robotics, modular robotics, amorphous computing, cellular automata, and synthetic biology. Bodies developed through morphogenetic engineering may take form through the birth and death of sessile elements, or through the rearrangement of motile ones. Design approaches also vary. Processes may be designed by hand and in an ad hoc fashion to illustrate particular hypotheses or principles. Designs may also be generated by hand in more systematic ways, for example through global- to-local compilation. Evolutionary algorithms are also popular for the automatic design of processes to achieve well-defined goals. Finally, a spectrum exists from biological morphogenesis with limited human control to fully artificial systems [44].

Embryological morphogenesis has inspired a variety of approaches to swarm intelligence [2, 7, 12, 17, 33, 34, 42]; our approach to artificial morphogenesis tends to adhere more closely to natural morphogenesis than most of these others do. Doursat, Sayama, and Michel [8] provide a rigorous taxonomy of morphogenetic engineering, including additional examples, and Oh et al. [36] provides a more recent review of similar work.

Morphogenetic engineering is a specific approach to the goal of *programmable matter*, that is, the ability to systematically control the properties and behavior of material systems at a fine level [14, 30], and artificial morphogenesis exhibits the properties of *active matter* [6, 35, 46]. Morphogenetic engineering also has some similarities to *amorphous computing* [1], but the artificial morphogenesis processes may begin with simple, structured preparations (see for example Sec. 5 below), and even if they do begin in an unorganized state, it is characteristic of the morphogenetic processes to quickly self-organize.

Although natural morphogenesis is a complex and intricate process— still incompletely understood—biologists have identified about twenty fundamental processes [10, pp. 158–9, 40]. Not all are applicable to artificial morphogenesis, and some (such as cell divisions) may be difficult to implement, nevertheless these processes provide an agenda for morphogenetic engineering, since they are in principle sufficient for assembling something as complex as an animal's body [23, 27–29]. Processes that cannot be directly implemented (such as cell division) may need to have their function accomplished by alternative means (e.g., providing components from an external source and moving them to the growth zone; see Sec. 5.1 for an example).

During biological morphogenesis, cells emit and receive chemical signals, migrate and differentiate in response to those signals, and participate in both reproduction (*cell proliferation*) and programmed cell death (*apoptosis*). In the process they emit and absorb molecules that serve both as communication media (*morphogens*) and as structural elements. Similarly, in our approach to artificial morphogenesis we distinguish *active components* and *passive components* [30]. Active components are microscopic agents (microrobots or genetically engineered micro-organisms) that can emit and respond to simple signals, implement simple (primarily analog) control processes, and move and transport other (active or passive) components. Passive components are all the rest, including signaling molecules and structural components. They do not move under their own power, but are moved by external forces (including Brownian motion and active components). In a typical artificial morphogenesis process, agents (active components) might transport and assemble passive components into a desired structure, or they might assemble themselves into the structure, as cells do in biological development.

2. Describing Artificial Morphogenesis with PDEs

The challenge of artificial morphogenesis is to control very large swarms of active components (microscopic agents) to interact with each other and with the passive components to assemble a desired structure. This requires an appropriate level of abstraction that allows the process to be described in sufficient detail for implementation without becoming obscured by details of a specific implementation technology. Here again we have taken our inspiration from biology, for biologists often use partial differential equations (PDEs) to describe the morphogenetic processes. At this level of abstraction developing tissues and diffusing signaling chemicals are described at a macroscopic level largely independent of individual cells. It also allows the methods of continuum mechanics to be applied, which are especially appropriate for embryological development, which takes place in the domain of "soft matter" (viscoelastic materials) [4, 10]. Therefore we are operating in the domain of well-understood mathematical methods in which continuum models are applied to macroscopic volumes of materials with a discrete molecular structure.

Scalability is a principal advantage of using PDEs for morphogenetic engineering. This is because a PDE, especially when used in continuum mechanics or fluid dynamics, treats a material as a phenomenological continuum; that is, it is treated as though it is composed of an infinite number of infinitesimal particles, which is a usable approximation of a material composed of a very large number of very small particles. Therefore, if our goal in artificial morphogenesis is to use a very large swarm of very small agents, PDEs take this goal to the continuum limit. Instead of worrying whether our algorithms will scale up to larger swarms of smaller agents, we have the complementary, but easier, task of scaling down from an infinite number of infinitesimal agents to a very large number of very small agents.

Another advantage of expressing morphogenetic processes in PDEs is that they are largely independent of agent size. This is because the processes are expressed in terms of *intensive quantities,* such as agent density, rather than *extensive quantities,* such as numbers of agents [24, 28, 29]. Therefore, an algorithm that produces structures of a particular size will continue to be correct even if the size of the agents is changed. (There are obviously limits to this size independence; the continuum approximation has to be good.)

In summary, using PDEs allows us to describe the behavior of *massive swarms,* by which we mean swarms that can be treated as a continuous mass. We treat massive swarms the same way tissues are treated in embryological morphogenesis and the same way fluids and solid masses are treated in continuum mechanics.

3. A Morphogenetic Programming Language

In order to test algorithms for the coordination of microrobot swarms for morphogenetic engineering, we have developed a morphogenetic programming notation based on mathematical notation for PDEs [22, 25–27, 29]. In order to facilitate simulations, this notation has been formalized into a morphogenetic programming language, tentatively named "Morphogen". We have a prototype implementation of Morphogen by means of a syntax macroprocessor, which translates Morphogen programs into MATLAB® or compatible GNU Octave [31]. This approach imposes some syntactic limitations on the Morphogen language, which would not be required with a conventional compiled implementation, but it permits rapid prototyping and experimentation with language features. There are two slightly different dialects of the language for describing either two-dimensional or three-dimensional morphogenetic systems. (A complete grammar for a previous version of Morphogen is published in MacLennan [21]; in this chapter we use Morphogen version 0.14.)

3.1 Substances

A morphogenetic program is organized into a number of *substances* with common properties; typical substances include diffusing morphogens and massive swarms of agents. Substances are similar to classes in object-oriented programming languages, in that a substance defines the common properties of an unlimited number of particular instances (called *bodies* in Morphogen, analogous to objects in object-oriented programming). Like subclasses in object-oriented programming, more specific substances may be derived from more general substances by specification of parameters that were unspecified in the more general substance, and by the addition of properties and behaviors in the derived substance. For example, we might have a substance representing a general diffusible substance with an unspecified diffusion rate, and one or more specific diffusible substances, with specified diffusion rates, derived from the general substance.

We distinguish *physical substances* and *controllable substances*, but this is more a matter of degree than kind, and applies primarily to active components. The idea is that the properties and behavior of physical substances are relatively fixed. For example, a particular chemical morphogen will have specific diffusion and decay rates within a given medium, and a particular agent swarm will be composed of agents, such as microrobots, with particular masses and specific sensors and actuators. Changing these properties entails making a new substance, new kinds of agents, and different hardware.

In contrast, a controllable substance has properties that are relatively easy to control. For example, a programmable agent will permit its sensors and actuators to be controlled so that it behaves in a desired way; its software can be changed. That is, a controllable substance is in some sense programmable, but there are degrees of programmability.

The current Morphogen language does not distinguish between physical and controllable substances, since controllability is a matter of degree and even economics. However, the distinction might be expressed in a substance hierarchy. For example, a substance definition might reflect the physical properties and behavior of a certain kind of microrobot, and then various substances derived from it could represent the microrobots programmed for distinct functions and behaviors.

This is an example of a simple substance definition (a diffusible morphogen), which illustrates its parts:

```
substance morphogen :
    scalar field C                  // concentration
    behavior :
        param d_C = 0.3             // diffusion constant
        param t_C = 10              // decay time constant
        D C = d_C * del^2 C - C/ t_C // diffusion & decay
```

As illustrated by this example, the syntactic extent of many Morphogen constructs is indicated by indenting. (This substance definition is terminated by a line indented less or equal to that of the word **substance**.) A substance definition has two parts: field declarations (preceding **behavior**) and behavioral specifications (following **behavior**).

3.2 Field Declaration

A substance is characterized by one or more continuous *fields* defined throughout the space. These may be scalar fields, such as concentration or density fields, or vector fields, such as velocity or flux fields. The preceding definition of "morphogen" illustrates the declaration of a scalar field called "C"; an example vector field declaration is:

vector field V

Multiple fields of the same type can be declared on indented lines, as illustrated by this example:

scalar fields :
   C // *swarm density*
   S // *magnitude of morphogen gradient*
vector fields :
   U // *morphogen gradient*
   V // *swarm velocity*

3.3 Behavior Specification

Fundamental to our approach is the use of partial differential equations to describe the behavior of massive swarms of microscopic agents. Such agents and the materials that they control will move continuously in time, but we can also simulate this behavior on ordinary computers in discrete time. Therefore we describe the behavior of substances by *change equations* which can be interpreted ambiguously as either ordinary PDEs or as temporal finite difference equations. The derivation rules of this calculus respect both interpretations. We write change equations with the notation $ÐX = F(X, Y,...)$, which means either the differential equation $\partial_t X = F(X, Y,...)$ or the difference equation $\Delta X/\Delta t = F(X, Y,...)$. A change equation such as

$$ÐC = d_C \nabla^2 C - C/t_C$$

is written as follows in the Morphogen language:

$$\mathbf{D}\ C = d\_C * \mathbf{del}\char`\^2\ C - C/t\_C$$

The Morphogen prototype implementation does not include a full expression parser, and so there are some notational concessions to allow expressions to be handled by the syntax macroprocessor. For the most part, scalars and fields (both scalar and vector) can be combined using standard arithmetic operators (+, −, ×, /). However, products or quotients of two scalar fields, products of a scalar field and a vector field, quotients of a vector field and a scalar field, and powers of scalar fields must be surrounded by square brackets, for example, [C * V]. Morphogen includes several vector operators, including the gradient **del** X, the Laplacian **del**^2 X, the divergence **div** X, and the pointwise $\mathcal{L}_2$ norm of a vector field, $||X||$.

Some change equations are stochastic, either to model indeterminacy and uncertainty in the physical systems, or to introduce randomness into the swarm control (e.g., to break deadlocks and symmetry). In order to respect both continuous- and discrete-time stochastic change equations, the morphogenetic programming notation interprets $ĐW^n$ to be an n-dimensional random vector field distributed $\mathcal{N}$ (0, 1) in each dimension; it represents n independent sources of randomness at each point in the (two- or three-dimensional) space [22, 26, 29]. For a $d \in \{2, 3\}$ dimensional morphogenetic process, we usually use $MĐW^n$, where M is a $d \times n$ matrix that weights and sums the random inputs into a d-dimensional vector field. M can be a $1 \times n$ matrix to generate a scalar field, or a scalar when $n \in \{1, d\}$. All of these cases are written [M **DW**^n] in the Morphogen language.

Control of morphogenetic swarms often requires them to change their behavior when some quantity exceeds a threshold or is within some range. To accommodate these requirements, Morphogen permits bracketed conditional factors (actually, Heaviside step functions), which are 0 or 1 depending on whether or not the condition is true. For example, the following describes a scalar field A in which any value greater than θ increases exponentially until it saturates at $A = 1$; values less than or equal to θ decay to zero (with time constant τ):

$$D A = [A > \text{theta}]\ A * (1 - A) - [A <= \text{theta}]\ A/\text{tau}$$

Sometimes the behavior of a field is described by several *partial change equations* distributed among several substance definitions. Partial change equations are written in either of these forms:

$$D \langle name \rangle\ +=\ \langle expr \rangle \langle newline \rangle$$
$$D \langle name \rangle\ -=\ \langle expr \rangle \langle newline \rangle$$

(This notation was first used in morphogenetic programming by Fleischer [9, p. 20].) A typical application of it would be describing the behavior of a morphogen. The definition of the morphogen substance might define its diffusion and decay with a partial change equation such as this:

$$D M\ +=\ d\_M * \textbf{del}\ ^2\ M - M/\text{tau}\_C$$

This is part of the definition of the morphogen as it is a physical property of the substance. Elsewhere in the program, in the definition of a substance representing the agent swarm (with swarm density S), another partial change equation could describe the emission of the morphogen at a rate k_M:

$$\mathbf{D}\,M \mathrel{+}= k\_M * S$$

This is part of the behavior of the swarm substance because the morphogen is produced by the agents constituting the swarm. It is often convenient to name additional scalars or fields within a substance behavior definition, which is accomplished by a statement of the form:

$$\textbf{let } \langle name \rangle = \langle expr \rangle \langle newline \rangle$$

(In the current prototype implementation, declared fields have global scope, but **let**-defined variables are local to the substances in which they are defined.)

The behavior of substances is determined also by certain fixed parameters, such as diffusion and decay rates. These are specified by a *parameter definition*, which has the syntax:

$$\textbf{param } \langle name \rangle = \langle expr \rangle \langle newline \rangle$$

Parameter definitions usually occur at the beginning of the **behavior** part of a substance definition, but can also be global as part of the **simulation parameters** (see Sec. 3.5 below). A sequence of parameter definitions can be indented under the word **params**.

The following behavioral rules [21] illustrate many of the Morphogen features we have discussed (and also line continuation indicated by " ... "):

```
params :
    v  = 1             // base swarm speed
    lambda = 0.03      // density regulation
    eps  = 1e−100      // minimum gradient norm
    k_W = 0.1          // degree of random motion
    k_P  = 30          // path deposition rate
    t_D  = 5           // delay period
let U = del A
let S = || U ||
let V = [(v * U)/(S+eps) ] − lambda * del [(C − 1)^2] ...
    + [k_W DW^2]
D C  = [ t>t_D ] −div [ C * V ]  // change in density
D P += [ t>t_D ] k_P * [ C * (1 −P ) ]  // path deposition
```

3.4 Bodies

The substance definitions of a morphogenetic program describes the general properties of the materials or substances (including massive agent swarms) that are involved in the process. A complete morphogenetic program also requires specification of the initial conditions, which in the case of morphogenesis means an initial preparation of the substances. Since the goal of morphogenesis is for complex structures to emerge by means of self-assembly, the initial preparation should be simple in structure: simple spatial arrangements of materials and of sources and sinks.

In our case, the initial conditions are specified by defining a small number of *bodies* belonging to defined substances. A body definition has a header followed by a sequence of (indented) initializations, for example:

> **body** Obstacles **of** path_material : // *place obstacles*
>   **for** (x, y) **within** 0.06 **of** (−0.1, 0.225) : P = 1
>   **for** (x, y) **within** 0.06 **of** (0.1, −0.225) : P = 1

Initializations currently have the following forms:

⟨initialization⟩	::=	**for** ⟨region⟩ : ⟨init⟩
	|	**for** ⟨region⟩ : ⟨newline⟩ ⟨init⟩∗ ⟨dedent⟩
⟨region⟩	::=	⟨expr⟩ < ⟨name⟩ < ⟨expr⟩, ⟨expr⟩ < ⟨name⟩ < ⟨expr⟩
	|	(⟨name⟩, ⟨name⟩) **within** ⟨expr⟩ **of** (⟨expr⟩, ⟨expr⟩)
⟨init⟩	::=	⟨name⟩ = ⟨expr⟩⟨newline⟩

(⟨dedent⟩ represents delimitation by an indent level less than or equal to the beginning of the construct.) This syntax is for the 2D version of Morphogen; the 3D version has similar initializations. The second kind of **for** definition permits multiple fields to be initialized in a region, e.g.:

body Start **of** swarm
    for −0.5 < x < 0.5, 0 < y < 0.1 :
        S = 1   // *initial swarm density*
        M = 0.0 5 // *initial morphogen concentration*

(Uninitialized regions are assumed to be zero.)

3.5 Simulation

The preceding morphogenetic programming constructs apply to simulations as well as to specifying real morphogenetic processes executed by physical microrobots. For simulation purposes, the Morphogen language provides additional facilities. The overall structure of a Morphogen program has the syntax:

> **morphogenetic program** ⟨name⟩ :
>   ⟨sim params⟩
>   ⟨substance⟩∗
>   ⟨body⟩∗

⟨visualization⟩
end program

The parameters for the simulation are followed by the substance definitions, which are followed by the body definitions, which are followed by visualization commands.

The block of simulation parameters has the syntax:

⟨sim params⟩   ::=   **simulation parameters** : ⟨sim par⟩* ⟨dedent⟩

⟨sim par⟩   ::=   **duration** = ⟨num⟩⟨newline⟩

             | **temporal resolution** = ⟨num⟩⟨newline⟩

             | **space** ⟨num⟩ < **x** < ⟨num⟩, ⟨num⟩ < **y** < ⟨num⟩⟨newline⟩

             | **spatial resolution** = ⟨num⟩⟨newline⟩

             | **save** ⟨name⟩⁺ **to** ⟨filename⟩⟨newline⟩

             | **load** ⟨name⟩⁺ **from** ⟨filename⟩⟨newline⟩

             | ⟨log params⟩

⟨log params⟩   ::=   **log params** ⟨name⟩ [, ⟨name⟩]* ⟨newline⟩

             | **log note** ⟨characters⟩ ⟨newline⟩

The **space** specification defines the (two- or three-dimensional) region in which the morphogenetic process takes place; **duration** determines how long (in simulated time) the simulation runs. The spatial and temporal resolutions of the simulation are defined by the **spatial resolution** and **temporal resolution** specifications. The **save** and **load** directives allow scalar and vector fields to be saved and restored. Finally, **log params** records the specified parameter values in a time-stamped text file, and **log note** puts a note in it.

Morphogen provides visualization commands to display fields in a variety of formats either during the simulation (**running**) or at its end (**final**):

⟨visualization⟩   ::=   **visualization** : ⟨command⟩⁺ ⟨dedent⟩

⟨command⟩   ::=   **display** ⟨time⟩⟨primitive⟩ **as** ⟨kind⟩ ⟨opt⟩

             | **make movie** ⟨filename⟩ **of** ⟨primitive⟩ **as** ⟨kind⟩ ⟨opt⟩

             | ⟨stability⟩

⟨time⟩   ::=   **running** | **final**

⟨kind⟩   ::=   {**mesh**|**contours**|**colors**} [**limits** (⟨expr⟩, ⟨expr⟩)]

             | **quivers** [⟨primitive⟩ **mesh**]

⟨opt⟩   ::=   ⟨characters⟩ ⟨newline⟩

For example, the following command displays a scalar field C as a heat map during the simulation:

> **display running** C **as colors**

The following displays a vector field V as an array of quivers on a 0.2 × 0.2 mesh at the completion of the simulation:

> **display final** V **as quivers** 0.2 **mesh**

A running display of a field can be converted to a movie, e.g.,

> **make movie** Cvid.mp4 **of** C **as contours**

Finally, Morphogen provides several commands for assessing the numerical stability of the simulation:

⟨stability⟩ ::= **report diffusion number for** ⟨primitive⟩
| **report Courant number for** ⟨primitive⟩
| **report Peclet number for** ⟨primitive⟩ **and** ⟨primitive⟩

4. Example: Routing Neural Pathways

Suppose we wanted to assemble a neuromorphic computer with a complexity comparable to a mammalian brain. Such a neurocomputer might have billions of artificial neurons, each with many thousands of connections, and it would be organized into functional regions, each comprising many millions of neurons, with remote regions connected by dense bundles of millions of neural fibers (artificial axons). In a developing mammalian brain, axons grow toward their destinations by following chemical signals (morphogens) that indicate way stations and the final destinations. The stepwise development of a morphogenetic program to solve this problem illustrates our approach to swarm intelligence.

4.1 Discrete Agents

We begin with a simple path routing process. An agent will be placed at the connection's origin (e.g., the originating point for a neural fiber), and a source of a diffusing morphogen (the attractant) will be placed at its destination. If the agent moves up the attractant gradient, then it will find its way to the destination, and it can create in its wake the path from the origin to the destination. However, we want to create many such paths, and so they cannot go in a straight line to their destinations, but must weave their way around already created paths. There are at least two methods to accomplish this. One is to have existing paths emit a repellant morphogen, which the agents try to avoid while seeking the attractant [23]. An alternative, simpler solution is to have the existing paths absorb or degrade the morphogens, so that they become sinks for it [24, 27]. Although both solutions work, in this case the simpler solution works better, since the concentration of diffusing morphogens decreases exponentially with distance from the source [21]. For a 3-dimensional diffusion-decay process $\dot{C} = D\nabla^2 C - C/\tau$, the steady-state concentration at a distance r from a point source with rate k is $C(r) = \dfrac{k}{4\pi Dr}\exp(-r/\sqrt{D\tau})$. Since the concentration of attractant varies significantly with distance from the destination, it is difficult, with the first approach, to properly balance it against the repellant throughout the space. This balance is achieved in the second approach because the existing paths absorb a fraction of the attractant in their vicinity.

Our goal, however, is not to have single neural fibers connecting (artificial) brain regions, but to have dense bundles of point-to-point connections. Therefore, rather than having a single agent follow the attractant gradient from the origin to the destination, we want a large swarm of agents to do so, creating in its wake a bundle of fibers that is tight but not too tight. Flocks of birds and schools of fish move in an organized way in compact groups, and so we have explored a modified flocking algorithm as a way of coordinating a swarm to lay down fiber bundles from an origin region to a destination region [23, 24]. This algorithm has been shown to scale from five agents (and hence five fibers per bundle) up to 5000 agents and hence 5000 fibers per bundle, with no change of parameters and a small number of errors (acceptable in neural computation): scaling over four orders of magnitude [24]. See Fig. 1.

4.2 Continuous Swarm

Next, we take the number of agents in the swarm to the continuum limit. The development and refinement of the morphogenetic algorithm is described in MacLennan [21]; here we summarize the final algorithm.

The path routing algorithm makes use of four substances: an attractant morphogen, the goal material marking the destination, the path material laid down by the agents, and the swarm substance, which is composed of agents. The *morphogen* substance is defined by a concentration field and a diffusion-decay equation, which describes its physical behavior:

```
substance morphogen :
      scalar field A                      // morphogen concentration
    behavior :
      param d_A = 0.03                    // diffusion constant
      param tau_A = 100                   // decay time constant
      D A += d_A * del^2 A − A/tau_A      // diffusion + decay
```

The change in A is described by a partial change equation because this reflects only its physical diffusion and decay; it is also emitted by the *goal* material and absorbed by the *path* material.

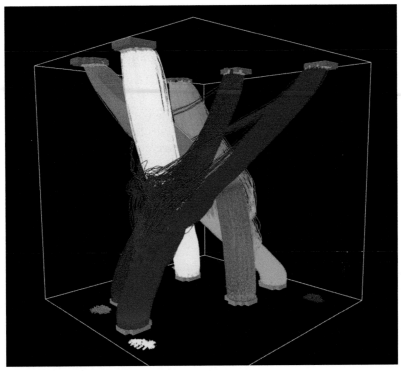

Fig. 1. Neural fiber bundle routing by modified flocking algorithm. There are five bundles, each comprising 5000 fibers, joining randomly selected origins and destinations on the lower and upper surfaces.

The *goal* material emits the attractant morphogen (and in this sense is active), but otherwise does not change:

```
substance  goal_material :
     scalar field G            // density of goal material
   behavior :
     param k_G = 100           // attractant release rate
     D G = 0                   // G field is fixed
     D A += k_G * [G*(1 –A ) ]  // goal emits attractant
```

The partial change equation for A describes its emission by the *goal* material up to saturation at $A = 1$ and therefore acts as a source term.

The *path* material P, which is laid down by the swarm, has two properties:

(1) It absorbs the attractant morphogen (so that new paths avoid existing paths), which is defined by a partial change equation for the morphogen, $ÐA -= PA/\tau_p$. Therefore the complete PDE for the morphogen is:

$$\dot{A} = d_A \nabla^2 A - A / \tau_A + k_G G(1 - A) - PA / \tau_p$$

(2) P increases its concentration autocatalytically if its concentration is above a threshold θ_P, and it decays if it is below the threshold, which is described by the partial change equation:

$$\text{Ð}P \mathrel{+}= a_P [P > \theta_P] P (1 - P) - [P \le \theta_P] P / t_P. \tag{1}$$

This autocatalytic behavior sharpens up the paths and ensures they have consistent density, which the equation drives to $P = 1$ (path present) or $P = 0$ (path absent). It is a partial equation because the swarm also deposits path material. The foregoing behavioral specifications are combined in the *path material* substance definition:

```
substance  path_material :
    scalar  field  P        // path  density
  behavior :
    params :
      a_P = 20              // autocatalytic  rate
      theta_P = 0.3         // autocatalytic  threshold
      theta_C = 0.02        // quorum  threshold
      t_P = 1               // path  decay  time  constant
      tau_P = 0.2           // attractant  absorption  time
    // autocatalysis :
    D  P = a_P * [ P>theta_P ] [ P * (1 -P) ]  - [ P<=theta_P ] P/ t_P
    D  A -= [ P * A ] / tau_P   // path  absorbs  attractant
```

Finally we define the *swarm* substance, which controls the mass of agents to lay down a path while approaching the goal and avoiding existing paths. The principal variable is the swarm density C, which is a real number reflecting the continuum approximation. The *swarm* substance also has a vector field $\mathbf{V}$, which defines the velocity of the swarm throughout space. The change in swarm density, then, is simply the negative divergence of the flux $C\,\mathbf{V}$:

$$\text{Ð}C = -[t > t_D]\ \mathrm{div}\ C\,\mathbf{V}.$$

The conditional factor $[t > t_D]$ prevents swarm movement until t_D time units have passed, which allows the morphogen to diffuse throughout the space.

The velocity field is a weighted combination of three influences: the normalized morphogen gradient, a density control term, and random (Brownian) motion to break symmetries. The normalized morphogen gradient (or gradient versor) $\mathbf{V}_1$ is $\nabla A / \|\nabla A\|$, but to avoid possible division by zero, we compute it: $\mathbf{V}_1 = \mathbf{U}/(\|\mathbf{U}\| + \varepsilon)$, where $\mathbf{U} = \nabla A$ and ε is a small number. To control the density, we define a potential function $(C - 1)^2$ that is minimized at the desired density $C = 1$. The density control velocity is the negative normalized gradient of this potential: $\mathbf{V}_2 = -\mathbf{W}/$ $(\|\mathbf{W}\| + \varepsilon)$, where $\mathbf{W} = \nabla[(C-1)^2]$. A regularization parameter λ controls the relative magnitude of these two velocity fields, $(1 - \lambda)\mathbf{V}_1 + \lambda\mathbf{V}_2$, which are combined with Brownian motion $k_W \eth W^2$ and an overall speed v to compute the final velocity field:[1]

$$\mathbf{V} = v[(1 - \lambda)\mathbf{V}_1 - \lambda\mathbf{V}_2 + k_W \eth W^2)]. \tag{2}$$

In Morphogen the substance definition begins like this:

```
substance swarm :
    scalar field  C              // swarm density
    vector  fields :
        U                        // morphogen gradient
        V                        // swarm  velocity
        W                        // density gradient
    behavior :
    params :
        v  = 1                   // base swarm speed
        lambda = 0 . 1           // density regulation
        eps  = 1e–100            // minimum  gradient norm
        k_W  = 0 . 3             // degree of random  motion
        t_D  = 5                 // time delay
    let U = del  A               // morphogen  gradient
    let W = del [ ( C– 1 ) ^ 2 ]   // density  gradient
    let V = v * ( ( 1 – lambda ) * [ U / ( ||U|| + eps ) ]   ...
          – lambda * [W/ ( ||W|| + eps ) ] + [ k_W DW^2 ] )
    D C = [ t>t_D ] –div [C * V]
```

Finally and most importantly the swarm deposits path material at a relative rate k_P but saturates at $P = 1$. Therefore the swarm contributes to the change in P by $\eth P += [t > t_D]k_P C (1 - P)$, which begins after t_D time units. This is expressed in Morphogen:

```
param  k_P = 30               // path  deposition  rate
D P += [ t>t_D]  k_P * [ C*(1 –P) ]  // path  deposition
```

To test the algorithm we can define some obstacles and an origin and destination for a new path. Everything we have discussed so far works for either a two- or three-dimensional morphogenetic system, but there are minor differences between a 2D and 3D simulation.

For a 2D simulation, the obstacles (representing previously generated paths) are just circular regions of path material. These can be created by a **body** definition such as shown in Section 3.4. For 3D simulations there are

---

[1] Alternately, the Brownian motion can be described by a diffusion term.

additional ways to create obstacles, or the algorithm itself can be used to generate multiple paths, as in the earlier simulations with discrete swarms.

To initialize the creation of a path, the swarm needs to be placed at the origin, and goal material needs to be placed at the destination. Here are typical body definitions:

```
body Cohort of swarm :
    for  −0.05 < x < 0.05 ,  −0.95 < y < − 0.9 :  C = 1

body Goal of  goal material :
    for  −0.05 < x < 0.05 ,  0.9 < y < 0.95 :  G = 1
```

A series of simulations exploring algorithm variants and parameter values is described in MacLennan [21]; here we present typical results in Fig. 2.

Figure 2a shows the result of a 2D simulation in which the origin was placed near the lower right corner and the destination was placed near the upper left. The nine disks represent previous paths, which are obstacles to be avoided. The new path finds its way around the obstacles, splitting to do so (similar to the bundles in Fig. 1). The degree of allowed splitting can be controlled by the swarm density regularization parameter λ (Eq. 2). The edges of the bundle are sharp and the density is constant due to the autocatalysis of path material (Eq. 1).

Figure 2b shows the result of a 3D simulation in which the origin was placed near the foreground and the destination was placed on the rear surface. The path found its way around four obstacles representing

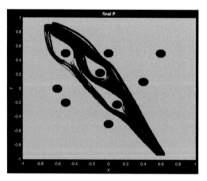
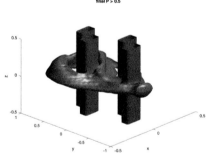

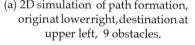

(a) 2D simulation of path formation,   (b) 3D simulation of path formation, origin in
origin at lower right, destination at   foreground, destination on back surface,
upper left, 9 obstacles.   four obstacles. The figure shows regions of
                                                             path density $P > 0.5$.

Fig. 2. Simulations of neural path formation. In both simulations it can be seen that a swarm sometimes splits to go around an obstacle. For more information, see MacLennan [21].

previously generated paths. The 2D and 3D morphogenetic programs are essentially identical.

5. Example: Body and Leg Segmentation

Our second example illustrates how a swarm of microscopic agents can be coordinated to produce a simple insect-like robot body with a segmented "spine" and segmented "legs." (A typical 2D result is shown in Fig. 3.) The algorithm uses a process inspired directly by embryological spinal development but it uses it in two different ways: to segment the robot's spine (similar to its function in vertebrate development) and to segment the robot's legs (which is not how the legs of insects' or other animals' are segmented). Thus it shows how natural morphogenetic processes can be redeployed for different purposes in artificial morphogenesis.

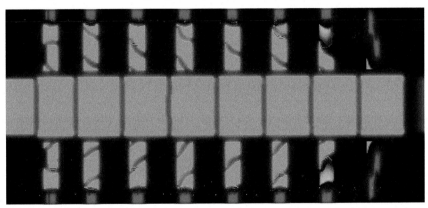

Fig. 3. Simulation of swarm assembly of insect-like robot body. The body is growing toward the right; the head segment is visible on the left, and the tailbud is at the extreme right but not shown. Between them, eight segments have been assembled, each with a pair of segmented legs (the last two pairs incomplete at this point in the simulation, $T = 60$).

The process in question is the *clock-and-wavefront model* of spinal somato-genesis, which was first proposed in 1976 and finally confirmed in 2008 [3, 5]. As the vertebrate embryo develops, a pacemaker within its tailbud periodically produces a pulse of chemicals, a segmentation morphogen. This chemical pulse propagates toward the head of the embryo, and is transmitted through the tissue, as each cell is stimulated by its neighbors to produce its own pulse. As each pulse passes through undifferentiated tissue, it causes the differentiation of one more somite or spinal segment. This takes place in a sensitive region defined by a low concentration of two morphogens: a *caudal morphogen*, which diffuses towards the head from the tailbud, and a *rostral morphogen*, which diffuses toward the tail from already differentiated segments. Therefore, the segments differentiate one by one from the head end toward the tail as the embryo grows. The ratio of the growth rate to the pacemaker frequency determines the length of the segments, while the product of the frequency and growth duration determines the number of segments (which is characteristic of a species).

5.1 Growth Process

Our morphogenetic program to assemble the robot body will make use of three different substances. The most important one is called *medium* and represents the mass of active components (the swarm) that does most of the work. This substance has a variety of properties, represented by scalar and vector fields; the most important is its density M; others will be introduced as required below. When it is part of a defined segment, *medium* is in a differentiated state, and the density of differentiated tissue is represented by a scalar field $S > 0$.

As in biology, differentiation means a change of internal state variables that affect the behavior and other properties of a substance. In this case, the differentiation into segment tissue could cause the agents to couple with their neighbors, resulting in a rigid structure, or to secrete some structural substance.

The second substance represents *terminal* tissue: either the tailbud of the developing spine or the "feet" of the developing legs; its principal property is a scalar field T, its density. Terminal tissue is composed of active components, and it is responsible both for growth processes and for generating the pacemaker signals. During the morphogenetic process, terminal tissue has a velocity: away from the head during spinal development, and away from the spinal axis during leg development.

The initial state of the morphogenetic process requires the preparation of two bodies: a head segment already in a differentiated state, and a tail segment adjacent to it. For a 2D simulation, this could be written as:

```
body  Head of  medium :
  for  0.01 < x < 1 ,  −0.5 < y < 0.5 :
     M = 1     // medium tissue
     S = 1     // segment tissue
     // other initialization
```

```
body  Tail  of  terminal :
  for  1 < x < 2 ,  −0.75 < y < 0.75 :
     T = 1    // initial tail bud
     // other initialization
```

As the terminal tissue moves away from medium tissue, the intervening space is filled with new, undifferentiated medium tissue. In biological development this is a result of cell proliferation, which expands the mass of undifferentiated tissue and pushes the tailbud further away from the head. In previous versions of our morphogenetic program, we have given the tail an initial velocity away from the head segment, and as it has moved it has produced undifferentiated tissue in its wake [23–25, 27, 28]. This approach makes sense if the agents are able to reproduce themselves, for example, if they are implemented by genetically engineered microorganisms. If the agents are microrobots, however, and are presumably unable to reproduce, then we need an alternative approach to growth. In this case the agents can be supplied from an external source and routed to the growth region. As tail tissue moves, it emits an attractant morphogen into the growth area (the growing gap between the existing body and the tail). Available agents from the external source follow the attractant gradient into the growth area until they reach a desired density ($M = 1$ in our case). Figure 4 shows a simulation of growth in progress.

This approach requires a source of free agents (particles of *medium*), and so we define a substance *source* to represent this process; it has a scalar field J representing the density of source material ($J \in \{0, 1\}$). The source provides agents at rate κ_J up to saturation ($M = 1$); therefore, as agents move out of the source regions, more will be introduced to maintain $M = 1$ density at the source. This provision of free agents takes place only so long as growth continues, which is represented by the condition $G > \vartheta_G$. Therefore the source of new agents is described $\text{Ð}M \mathrel{+}= [G > \vartheta_G]\kappa_J J (1 - M)$. The *source* substance is defined in this manner in Morphogen:

```
substance  source :
    scalar field J          // source density
  behavior :
    param  k_J = 1          // input rate
    D M += [G > theta_G]  k_J * [ J * ( 1 − M ) ]
```

Source regions must be prepared as part of the initialization of the morphogenetic process. For example, the simulation in Fig. 4 uses the following sources:

```
body Agent Sources of source :
    for 5 < x < 6, 1.8 < y < 1.95 : J = 1       // upper source
    for 5 < x < 6, −1.95 < y < − 1.8 : J = 1    // lower source
```

Terminal tissue has the following behavior. Let **u** be a unit vector field pointing in the direction of desired growth, and let r be the desired growth rate, which is the same as the speed of terminal tissue movement. Then the change in terminal concentration is the negative divergence of its flux:

$$\text{Ð}T = - \text{div}(T\, \mathbf{u}[G > \vartheta_G]r).$$

The condition $G > \vartheta_G$ holds so long as growth continues; when it fails, the movement rate is effectively zero and the terminal tissue stops moving. Figure 4a shows the massive swarm of terminal agents moving to the right.

Terminal tissue emits a growth attractant, with concentration H, up to saturation, which diffuses and decays at specified rates:

$$\eth H \mathrel{+}= \kappa_H T (1 - H),$$
$$\eth H \mathrel{+}= D_H \nabla^2 H - H/\tau_H.$$

Figure 4b shows the diffusion of the attractant from the terminal tissue.

The swarm of agents M from the sources follows the attractant gradient ∇H at a speed controlled by v and aggregating up to a maximum density of $M = 1$:

$$\eth M = -v[M < 1] \operatorname{div}[M \nabla H]. \tag{3}$$

Once the agents have assembled in the spine (where $M = 1$) they do not move again. Figure 4c shows the growth of undifferentiated spinal tissue in progress.[2]

5.2 Spine Assembly

The foregoing explains the growth of undifferentiated spinal tissue; we turn now to the clock-and-wavefront process, which causes the spine to differentiate into segments of specified number and size.

The concentrations of the caudal and rostral morphogens are represented by scalar fields C and R, respectively (Fig. 5). The rostral morphogen is produced by already differentiated segment tissue (represented by segment density $S > 0$), from which it diffuses and decays:

$$\eth R = \kappa_R S(1 - R) + D_R \nabla^2 R - R/\tau_R. \tag{4}$$

The caudal morphogen concentration C is considered a property of the *medium* substance, and its diffusion and decay are physical properties defined by a partial equation in the behavior of *medium*:

$$\eth C \mathrel{+}= D_C \nabla^2 C - C/\tau_C.$$

The caudal morphogen is produced from terminal tissue (i.e., tissue with terminal density $T > 0$), which is part of the behavior of the *terminal* substance:

$$\eth C \mathrel{+}= \kappa_C T (1 - C).$$

The segmentation signal α is produced periodically in the tailbud and is actively propagated forward through the *medium* tissue (Fig. 6). The segmentation morphogen is a property of *medium*, which defines its physical diffusion and decay rates:

---

[2] Other simulation results in this chapter use the original growth process.

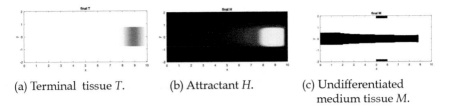

(a) Terminal tissue T.　　(b) Attractant H.　　(c) Undifferentiated medium tissue M.

Fig. 4. Simulation of spine growth process in progress ($t = 30$). (a) Density T of terminal tissue (moving to right). (b) Concentration H of attractant emitted by terminal tissue. (c) Density M of assembled undifferentiated tissue. Agents are introduced from the sources along the upper and lower edges and migrate up the attractant gradient to assemble into the spine.

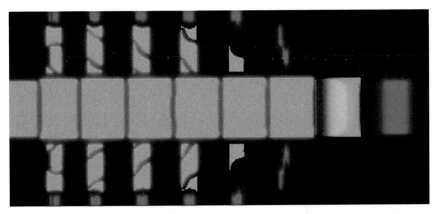

Fig. 5. Diffusion of C and R morphogens after seven new spinal segments have differentiated (tan color). Blue color at far right represents C (caudal morphogen) diffusing from the tailbud. Green color around the segments represents R (rostral morphogen) diffusing from differentiated segments. The α wave visible between the sixth and seventh segments has just differentiated the seventh segment (which is not yet emitting R). The developing legs are also emitting C and R morphogens.

$$\partial\alpha + = D_\alpha \nabla^2 \alpha - \alpha/\tau_\alpha.$$

Terminal tissue ($T > 0$) in the tailbud produces a pulse of α so long as growth is continuing (represented by $G > \vartheta_G$) and the tail's pacemakers are in the correct phase (represented by $K > \vartheta_K$):

$$\partial\alpha + = \kappa_\psi [G > \vartheta_G \wedge K > \vartheta_K]T(1 - \alpha).$$

This partial equation is an extension of α behavior in the definition of the *terminal* substance. (Additional detail on the pacemaker and growth duration can be found in MacLennan [24] and Sec. 5.3 below.)

The *medium* substance also produces α when it is sufficiently stimulated by it ($\alpha > \vartheta_\alpha$), which allows it to propagate a wave of α stimulation. After it

Fig. 6. Propagation of α morphogen. Two successive waves are propagating leftward toward the head. The decaying α pulse in the tailbud is visible at the right-hand end.

produces an α pulse, the tissue enters a refractory period (represented by $\rho \geq \vartheta_\rho$), which ensures that the wave is unidirectional (towards the head). The variable ϕ (which is only transiently nonzero) represents the density of *medium* tissue in this sensitive and stimulated state, and is used both to produce a pulse of α and to start the refractory timer:

$$\phi = [\alpha > \vartheta\alpha \wedge \rho < \vartheta_\rho]M, \tag{5}$$
$$Ð\alpha \mathrel{+}= \kappa_\phi \phi\,(1 - \alpha), \tag{6}$$
$$Ð\rho = \phi - \rho/\tau\,. \tag{7}$$

The actual differentiation of a region of *medium* tissue into a segment ($S > 0$) takes place when a sufficiently strong ($\alpha > \alpha_{lwb}$) segmentation wave passes through a region with sufficiently low rostral ($R < R_{upb}$) and caudal ($C < C_{upb}$) morphogen concentrations. This causes a rapid c_ς $(1 - S)$ increase in differentiated tissue, which then triggers autocatalytic differentiation $\kappa_S S(1 - S)$ until it is complete ($S = 1$):

$$ÐS = \kappa_S S(1 - S) + [\alpha > \alpha_{lwb} \wedge R < R_{upb} \wedge C < C_{upb}]c_\varsigma\,(1 - S)$$

$$= (\kappa_S S + [\alpha > \alpha_{lwb} \wedge R < R_{upb} \wedge C < C_{upb}]c_\varsigma)(1 - S).$$

This can be written as follows in the current version of the Morphogen language:

let chi $=$ [alpha > alpha_lwb] [R < R_upb] [C < C_upb] c_seg
D S $=$ [(chi + k_S * S) * (1 - S)]

There is a small gap between the differentiated segments because segmentation takes place in a region where the rostral concentration R has dropped to R_{upb} (see Figs. 3 and 5). To determine the width g of the gap, set the threshold R_{upb} equal to the concentration at a distance g from the anterior differentiated tissue: in a 2D space,

$$R_{upb} = \frac{\kappa_R}{2\pi D_R} K_0\big(g/\sqrt{D_R \tau_R}\,\big), \text{ where } K_0 \text{ is a modified Bessel function of}$$

the second kind (Basset function). Figure 7 shows a 3D simulation of segmentation using the same morphogenetic program and parameters as the 2D simulation.

final S > 0.5

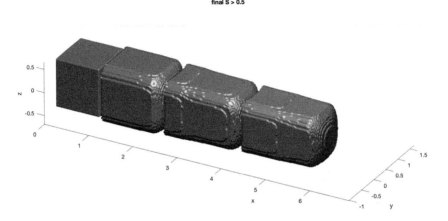

Fig. 7. Three-dimensional simulation of segmentation. Three new segments posterior to the head have been assembled. The figure shows regions in which $S > 0.5$.

5.3 Leg Assembly

The process described so far is analogous to the somitogenesis or spinal segmentation that takes place during vertebrate embryological development. Now however we exploit the process for a different morphogenetic task: the generation of a pair of segmented legs on each spinal segment (e.g. Fig. 3).

The first task is to control the placement of the legs on the spinal segments; in the case of a 2D simulation, the only variable is their location along the length of a segment. This can be determined by the relative concentration of morphogens that diffuse from the anterior and posterior ends of each segment (Fig. 8), but this requires distinguishing the anterior and posterior ends of each segment: a symmetry breaking process.

The required information is available when the segmentation α wave passes through the tissue, since at that time the more anterior tissue is emitting the rostral morphogen R, and the posterior tissue is emitting the caudal morphogen C. Therefore, the anterior region of a segment further differentiates into anterior tissue (with density A) when the segmentation wave passes through ($\alpha > \alpha_{lwb}$) and the rostral morphogen is in the correct range ($0.5R_{upb} > R > 0.25R_{upb}$ in this simulation), which triggers an autocatalytic differentiation process that goes to completion:

$$\mathcal{D}A = \kappa_A SA(1 - A) + [0.5R_{upb} > R > 0.25R_{upb} \wedge \alpha > \alpha_{lwb}]c_A - A/\tau_A.$$

Here, we expect R to reflect rostral morphogen diffusing from segments anterior to the currently differentiating segment. However, the differentiation of A tissue takes place simultaneously with the differentiation of S tissue (as α passes through), which begins immediately to emit R morphogen. This will cause the entire segment to differentiate into A tissue, and to prevent this we introduce a state variable B that temporarily blocks production of the rostral morphogen until anterior tissue differentiation is complete. It is set by the segmentation impulse (Eq. 5) and then decays exponentially:

$$ÐB = c_B\phi - B/\tau_B + [B > 1](1 - B)/t_B.$$

We choose the time constant τ_B to allow A differentiation to complete. We replace Eq. 4 for the production of R with:

$$ÐR = \kappa_R [B < \theta_B] S (1 - R) + D_R \nabla^2 R - R/\tau_R,$$

which blocks production until B has decayed below θ_B.

Similarly, the α wave and caudal morphogen in an appropriate range $(0.95C_{upb} > C > 0.8C_{upb})$ trigger autocatalytic differentiation of posterior border tissue:

$$ÐP = \kappa_P SP (1 - P) + [0.95C_{upb} > C > 0.8C_{upb} \wedge \alpha > \alpha_{lwb}]c_P - P/\tau_P.$$

(Emission of caudal morphogen does not have to be blocked since it comes from the tailbud.)

Once the anterior and posterior border tissues have differentiated, they can do their job of emitting anterior and posterior border morphogens, which provide a reference frame for determining position along each segment:

$$Ða = [A > \vartheta_A] \kappa_a S(1 - a) + D_a \nabla^2 a - a/\tau_a,$$
$$Ðp = [P > \vartheta_P]\kappa_p S(1 - p) + Dp \nabla^2 p - p/\tau_p.$$

The thresholds $(A > \vartheta_A, P > \vartheta_P)$ ensure that only well-differentiated tissue produces these morphogens.

Our plan is for the legs to be assembled on *imaginal tissue* at the correct location on each segment. The anterior/posterior position is defined by anterior and posterior border morphogens in the correct ranges $(a_{upb} > a > a_{lwb}, p_{upb} > p > p_{lwb})$, but these morphogens diffuse throughout the spinal tissue and we want to ensure that imaginal tissue differentiates only on the surface of the spine. Moreover, this notion of being "on the surface" should be independent of the agent size (which is in fact infinitesimal in our continuum model), so that the morphogenetic algorithm is scale-invariant. Taking a cue from the natural world, in which bacteria and other organisms do *quorum sensing* to determine if their population density is sufficient for some purpose, we have our agents estimate their local population density. If it is near its maximum ($M \approx 1$ in our case), then an agent knows it is in the interior; if it is near its minimum ($M \approx 0$) then the agent is relatively isolated. If however the local density is $M \approx 0.5$, then the agent knows it is near the surface; how large $|M - 0.5|$ is allowed to be and will determine the effective thickness of the surface layer.

This strategy requires that an agent be able to determine its local population density. There are many ways to do this, but perhaps the most practical is for the agents to emit a slowly diffusing, rapidly decaying morphogen N, which serves to broadcast density information over a short range. The local morphogen concentration becomes a surrogate for population density, more precisely, for the density convolved with a smoothing kernel. (See Sec. 6 for more on implementation of this strategy.) Morphogen code such as the following can be used to define a scalar field E that represents closeness to the surface (or in the case of 2D simulations, the edge).

D N = k_N * [M * (1 – N)] + D_N * **del**^2 N – N/tau_N
let E = [N > 0.15] [N < 0.2] S // *edge marker*

Then the autocatalytic differentiation of imaginal tissue can be triggered by a combination of correct morphogen ranges and being near to the surface:

$$Đ I = [a_{\text{upb}} > a > a_{\text{lwb}} \wedge p_{\text{upb}} > p > p_{\text{lwb}}]ES(1 - I). \qquad (8)$$

This equation is written as follows in Morphogen:

D I = [a_upb > a] [a > a_lwb] [p_upb > p] [p > p_lwb]...
   [E * [S * (1 – I)]]

Once the imaginal tissue has fully differentiated, the next step is to initialize the clock-and-wavefront process to grow and segment the legs. This is accomplished by having the imaginal tissue differentiate into terminal tissue with a velocity vector directed outward from the spine; thus the imaginal tissue becomes the "foot" of a future leg. We have used the rapid differentiation of imaginal tissue ($ĐI > \vartheta_{\text{D}I}$) to trigger this transformation. The velocity vector is directed down the S gradient: $-\nabla S/ \|\nabla S\|$. The leg grows just like the spine grows, recruiting free agents from the sources to fill in the gap between the developing leg and the outward moving foot. The exact same morphogens (H, R, C, α, a, p) and tissue types (T, S, A, P) are used to control leg segmentation as to control spinal segmentation.

5.4 Termination

There are two problems with using the same morphogenetic process for the spine and for the legs. The first is that with the same parameters, the number and size of the segments will be the same in the spine and the legs, which might not be what we want. The second is that, in the absence of a mechanism to prevent it, the leg segments will develop their own imaginal tissue, from which "leglets" will grow, and so on, in a fractal manner, which is not our goal. Therefore we must consider the termination of morphogenetic processes, which in fact is an interesting problem in embryology. How big should limbs and organs get? What limits growth processes in normal development?

In our morphogenetic program we have adopted a simple mechanism. In several of the equations above, we have seen that a process continues only so long as a condition $G > \vartheta_G$ is satisfied. Here G is a scalar property of terminal tissue, which decays exponentially: $Ð_G = -G/\tau_G$, where τ_G is the growth time constant. If the initial value of G is G_0, then the ϑ_G threshold will be reached at time $t = \tau_G \ln(G_0/\vartheta_G)$. If v is the pacemaker frequency, then $\lfloor v\tau_G \ln(G_0/\vartheta_G) \rfloor$ complete segments will be generated. The length of the segments will be v/ν, where v is the growth rate (the speed of terminal tissue movement). There are a number of parameters here that could be controlled, but if we want the same processes operating in spine and leg segmentation, then probably the simplest are the initial timer value and pacemaker frequency. For spinal segments of length λ_S, we set the pacemaker frequency to $\nu_S = v/\lambda_S$. For N_S spinal segments, we initialize the timer to

$$G_S = \vartheta_G \exp\left(\frac{N_S}{\nu_S \tau_G}\right)$$

See Fig. 9 for the effect of ν_S on spinal segment length and number. The equations for N_L leg segments of length λ_L are analogous.

As mentioned above, rapid differentiation of imaginal tissue can be used as a transient impulse to initialize leg growth; we define it $\zeta = [Ð I > \vartheta_{DI}]c_L$. This impulse can be used to trigger conversion of imaginal tissues to terminal tissue and to initialize the leg timers and pacemaker frequencies:

$$Ð T + = \kappa_L IT(1 - T) + \zeta, \tag{9}$$

$$Ð G + = \kappa_{GL} I\zeta(G_L - G), \tag{10}$$

$$Ð \nu + = \kappa\nu I\zeta(\nu_L - \nu). \tag{11}$$

κ_{GL} and κ_ν are large so they reset the parameters while $\zeta > 0$. Agents in imaginal tissue also reorient their velocity vectors down the S gradient, that is, outward from the spine:

$$Ð \mathbf{u} + = I\kappa_u (-\nabla S/||\nabla S|| - \mathbf{u}).$$

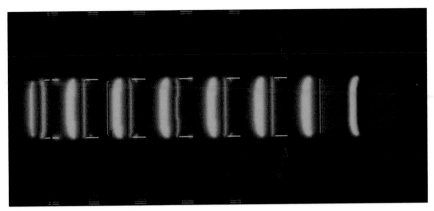

Fig. 8: Diffusion of a and p morphogens from anterior and posterior tissue, respectively, in each differentiated segment. Orange color represents a morphogen diffusing from differentiated anterior tissue, and green color represents p morphogen diffusing from differentiated posterior tissue. White lines on the spinal surface are imaginal tissue I from which the legs grow. (Two α waves are also visible.)

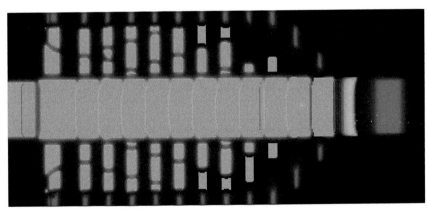

Fig. 9. Segmentation with higher pacemaker frequency. This simulation has twice the pacemaker frequency v_S as the simulation shown in Fig. 5. Both are shown at the same simulation time, but this has assembled 12 segments (compared to six in the previous simulation) with ten pairs of legs compared to five previously; the segments are also approximately half the length as in the previous simulation. In this simulation $v_S = 1/\pi$, the previous had $v_S = 1/(2\pi)$.

There are several ways to prevent the formation of fractal "leglets" on the legs; the most direct is to block the differentiation of imaginal tissue, as given by Eq. 8. The problem is that this equation is part of the behavior of *medium* tissue, which is the same in the spine and legs, since its constituent agents come from a common source. Leg terminal tissue, however, develops from imaginal tissue, but tail terminal tissue does not, which provides a basis for distinguishing the legs from the spine. Therefore we introduce a variable L representing the density of tissue in the leg state. When the differentiation of imaginal tissue triggers the differentiation of leg terminal tissue (Eq. 9), this tissue can simultaneously differentiate into the L state, which differentiates it from tail terminal tissue: $\mathrm{D}L \mathrel{+}= \kappa_L\, I\, L(1 - L) + \zeta$. As *medium* particles arrive from the sources and assemble themselves between the foot terminal tissue and the growing legs, they must also inherit the L property. This can be accomplished by having L tissue emit a short-range morphogen μ that causes developing *medium* tissue to inherit the L property. Therefore, if the concentration of μ in *medium* tissue is above a threshold (meaning that it is near "foot" terminal tissue) and it is not spinal tissue, then it enters the L state:

$$\mathrm{D}L \mathrel{+}= \kappa_\zeta\,[\mu > \vartheta_\mu \wedge S < 0.5](1 - L).$$

Then we change imaginal tissue differentiation (Eq. 8) so that it takes place only in non-leg tissue ($L \approx 0$):

$$\mathrm{D}I = [a_{\mathrm{upb}} > a > a_{\mathrm{lwb}} \wedge p_{\mathrm{upb}} > p > p_{\mathrm{lwb}} \wedge L < \vartheta_L] E S (1 - I).$$

Finally, note that the spinal tissue between the last segment and the tailbud is undifferentiated because the caudal morphogen concentration C is too high. Therefore it does not develop legs and forms the tail proper between the leg-bearing segments and the tailbud. To determine the tail's length λ, set the caudal morphogen concentration at a distance λ in front of the tailbud to the caudal morphogen threshold for segmentation: in 2D, $C_{\mathrm{upb}} = \dfrac{\kappa_C}{2\pi D_C} K_0(\lambda / \sqrt{D_C \tau_C})$. Also, as can be seen in Figs. 3, 7, and 9, the first spinal segment posterior to the head can have a different length compared to the other segments; this is because its length depends on the initial phase of the pacemaker.

6. Toward a Morphogen Compiler

6.1 Agents as Substantial Particles

Our approach to morphogenetic engineering treats agent swarms and other substances from the perspective of continuum mechanics. That is, although our swarms and tissues are composed of finite numbers of discrete elements of finite size, our intended application is to very large numbers of very small elements, and so the continuum approximation is useful. From the continuum mechanics perspective, tissues and massive swarms are infinitely divisible into infinitesimal *material points* or *particles* representing differential volume elements of continua [24]. Therefore, a material point or particle does not represent a single agent (or molecule or other active or passive component), but a very tiny volume containing many such physical elements. As a consequence, thinking of the individual behavior of particles in such a continuum is helpful in understanding how the individual agents should behave, but it is not sufficient to equate particles and agents. For example, a particle might have a distribution of orientations or velocities, or be in a mixture of differentiation states, whereas individual agents have definite orientations, velocities, and differentiation states [22, 26, 29].

We have seen how PDEs can be used to describe the motion, differentiation, and other behavior of swarms of microscopic agents to assemble complex shapes, but these PDEs cannot be used directly to control the agents. These PDEs have been expressed in an *Eulerian* (spatial) reference frame, in which the derivatives are relative to fixed spatial locations and describe the changes of variables at those locations as particles flow through them. For example, we may be concerned with the change in temperature $q(t, \mathbf{p})$ at a particular location $\mathbf{p}$, that is, $\partial q(t, \mathbf{p})/\partial t$, as particles with different temperatures flow through it. More relevant for the control of agent swarms is a *Lagrangian* (material) reference frame, in which derivatives are relative to fixed particles and describe how the properties of those particles change as they move through space. Thus we might be more concerned with the temperature $Q(t, P)$ of a particle P and how it changes as that particle flows through space, $\partial Q(t, P)/\partial t$; the latter is called the *material* or *substantial* derivative and is commonly written DQ/Dt.

Since a particle's position is a function of its velocity $\mathbf{v}$, the material derivative can be expanded by the chain rule:

$$\frac{DQ}{Dt} = \frac{\partial q}{\partial t} + \mathbf{v} \cdot \nabla q.$$

For a vector property of particles, the formula is analogous:

$$\frac{D\mathbf{Q}}{Dt} = \frac{\partial \mathbf{q}}{\partial t} + \mathbf{v} \cdot \nabla \mathbf{q}$$

where $\nabla \mathbf{q}$ is a second-order tensor. (In a Cartesian coordinate frame $(\nabla \mathbf{q})_{jk} = \partial q_j / \partial x_k$ and $\mathbf{v} \cdot \nabla \mathbf{q} = \mathbf{v}^T \nabla \mathbf{q}$.)

6.2 Morphogen-based SPH Control

Morphogen describes continuous fields of infinite numbers of infinitesimal agents. A compiler for the Morphogen language would produce code for a given numbers of agents of given finite size, and would thus realize the scalability potential of the language's PDE approach. Such a compiler would be an example of what some authors have termed "global-to-local

(GTL)" compilation [15]. GTL compilation seeks to turn on its head the usual challenge of emergence: It seeks not to predict collective behavior from simple rules for individuals, but to derive such rules based on desired collective behavior. In this section we explore a novel, embodied variation of smoothed particle hydrodynamics (SPH) swarm robotic control as a partial basis for a GTL compiler for Morphogen.

SPH itself is a meshfree Lagrangian numerical method used in physics simulations and was introduced by Gingold and Monaghan [13] and Lucy [20]. SPH was first proposed as a method for controlling robot swarms by Perkinson and Shafai [38]. In SPH robotic literature to date, researchers have used SPH to cause swarms to emulate fluids or bodies much like fluids (e.g., [11, 32, 39, 41]). Such use is in keeping with the typical use of SPH in physics to simulate fluids. However, we believe the true potential of SPH in robotic control lies in its ability to simulate non-physical PDEs. A step toward this goal was taken by Pac, Erkmen, and Erkmen [37], who design swarms that emulate fluids, but with non-physical variation over time in fluid parameters (see Tilki and Erkmen [45] for a more recent paper continuing this work). Our work goes further, applying SPH to Morphogen programs that do not resemble physical fluid behavior. This is one key contribution that our work makes with respect to the SPH robotic control literature, which we address further in Sec. 6.8. However, we first discuss our other key contribution, a proposal for overcoming a communications challenge posed by traditional SPH robotics.

6.3 Natural Smoothing Functions

If an agent can produce a physical field around itself having certain properties, then implicit in this field is what we term a *natural smoothing function* (NSF). We will refer to such physical fields as "NSF fields" to distinguish them from the potentially more abstract fields which NSF fields may be used to implement, as described in the next section. The NSF field around a given agent must satisfy the following:

- The strength or intensity (we will say *value*) of the NSF field must decrease smoothly and monotonically with distance from the agent.
- The NSF field's value must approach zero with distance, and quickly enough that the integral of value over all space converges.
- The NSF field must be roughly symmetrical around the agent.
- The agent must have control over the amplitude of the NSF field independently of its shape.

Finally, the NSF fields of different agents must sum naturally in the environment, and each agent must be able to sense locally both the value and gradient of the summed NSF field. If these requirements are met, then groups of such agents can implement SPH robotic control. The first three of these requirements are derived from general requirements on SPH smoothing functions; see Liu and Liu [19] for a discussion. Another key requirement is that the integral of the smoothing function be unity, which we address in the next section.

In later sections, we explore one possibility in depth: that of agents that release and detect dilute substances that diffuse and degrade in a surrounding aqueous medium. This scenario is inspired by the cells of a developing embryo as it undergoes morphogenesis. We refer to the smoothing functions implicit in this mechanism as "morphogen smoothing functions" (MSF). A second possibility, not discussed further in the present work, is that of agents producing and sensing electromagnetic or acoustic NSF fields distinguished by frequency. Before addressing MSFs specifically, we discuss the NSF framework generally.

6.4 Incorporating SPH Estimates into the NSF Framework

In contrast to the term "NSF field," we use the term "field" by itself to describe more general, and potentially abstract, entities. These are the fields which traditional SPH seeks to simulate and which SPH swarm robotic control seeks to generate. Examples in the context of artificial morphogenesis are density, velocity, or differentiation state of a tissue or agent continuum.

A diversity of SPH-based estimates of fields and their derivatives exist. In general, an SPH estimate at a point is the sum over neighboring particles of a smoothing function, or one of its derivatives, multiplied by some expression involving various spatially-varying quantities. Each of these quantities (e.g., mass, density, or velocity) is associated either with the location of the desired estimate or with the location of a given neighboring particle. Some SPH estimates require only quantities associated with the locations of neighboring particles, and such forms permit ready incorporation into the NSF framework. The most important example is the estimate of a field itself at location x:

$$\langle f(x) \rangle = \sum_j \frac{m_j}{\rho_j} f(x_j) W_j(x) \tag{12}$$

where m_j, ρ_j, and $f(x_j)$ are the mass of, local density at, and field value at, each neighboring particle j.

We can perform this estimate in the NSF framework if we assume that each agent can produce an NSF field as described in Sec. 6.3. If an agent j controls the amplitude of its NSF field so that it has unity integral, then the measurement by an agent i of the value of the NSF field produced by agent j is equivalent to the evaluation of $W_j(x_i)$, where $W_j(x)$ is a smoothing function whose shape is determined by the physical nature of the produced NSF field. If instead agent j controls the amplitude of its NSF field so that its integral is $\frac{m_j}{\rho_j} f(x_j)$, then the measurement by agent i of agent j's NSF field is the entire term $\frac{m_j}{\rho_j} f(x_j)W_j(x_i)$. Because the NSF fields of each neighbor j sum naturally in the environment, the total NSF field value measured locally by agent i is the desired local estimate of a field $\langle f(x_i)\rangle$. (This procedure requires agents to estimate their local densities, which they can do through a similar NSF estimate based on $\langle \rho(x_i)\rangle = \sum_j m_j W_j(x_i).$)

Because Eq. 12 is valid at points between particles as well as at particles, a spatial derivative of the field of SPH estimates is an estimate of that derivative in the corresponding field function. For example,

$$\langle \nabla f(x_i)\rangle = \nabla_{x_i} \sum_j \frac{m_j}{\rho_j} f(x_j)W_j(x_i) \qquad (13)$$

In traditional SPH, where a separate calculation is needed for an SPH estimate at each point, the above fact is not directly useful. Instead, much of the value of traditional SPH comes from the ability to rearrange expressions so that derivatives are applied only to the smoothing function and can be taken analytically. For example,

$$\langle \nabla f(x_i)\rangle = \sum_j \frac{m_j}{\rho_j} f(x_j)\nabla_{x_i} W_j(x_i)$$

By contrast, in the NSF framework, SPH estimates are physically embodied by the total NSF field at a given point. Therefore, if technologically feasible, an agent can make an SPH estimate of a spatial derivative by sensing that derivative in the local NSF field. In particular, we assume agents can sense NSF field gradients locally, for example using differences between multiple sensors at their boundary, such that Eq. 13 is directly useful.

Where such direct estimates are not practical, it may be possible to manipulate other methods from the SPH literature into forms compatible with the NSF framework. The Laplacian operator is an important example. Huang et. al. [16] derive an SPH form of the Laplacian operator requiring no derivatives. In two dimensions:

$$\langle \Delta f(x_i)\rangle = \frac{2}{\alpha}\sum_j (f(x_j)- f(x_i))W_j(x_i)\frac{m_j}{\rho_j},$$

where α is a constant that depends on an integral relating to a given smoothing function. Although it is not possible in the NSF framework to take the pairwise difference as shown, we can rearrange and note that $\sum_j W_j(x_i)\frac{m_j}{\rho_j}$ is just an SPH estimate of unity, and so can be replaced by 1.

This gives us a form we can calculate in the NSF framework:

$$\langle \Delta f(x_i) \rangle = \frac{2}{\alpha} \left(\sum_j \frac{m_j}{\rho_j} f(x_j) W_j(x_i) - \sum_j \frac{m_j}{\rho_j} f(x_i) W_j(x_i) \right)$$

$$= \frac{2}{\alpha} \left(\sum_j \frac{m_j}{\rho_j} f(x_j) W_j(x_i) - 1 \cdot f(x_i) \right)$$

$$= \frac{2}{\alpha} \left(\langle f(x_i) \rangle - f(x_i) \right) \tag{14}$$

6.5 Benefits of NSF Framework

NSF-based SPH robotic control addresses communication challenges and inefficiencies in traditional SPH robotic control. Traditional methods require each agent, for each SPH estimate it makes, to communicate separately with $O(n)$ other agents, where n is the number of agents in its neighborhood. In contrast, NSF-based agents need only sense an NSF field value locally, a $O(1)$ operation. In establishing separate data links with its neighbors, a traditional agent must avoid communication interference with its neighbors, all of which may be simultaneously trying to establish links with *their* neighbors. By contrast, NSF-based agents communicate implicitly by sensing an NSF field locally and manipulating their contribution to it; therefore, they do not require any explicit data links or even a concept of having individual neighbors.

We can view these advantages over traditional SPH control as the benefits of embracing an embodied approach to artificial morphogenesis. In the agent-to-agent communication of traditional SPH control, information is an abstraction atop the physical fields chosen as communication media. Overcoming communication interference is thus a struggle against the way these fields interact in space. By contrast, the NSF framework works with, rather than against, the interaction (as well as diminution with distance) of the physical NSF fields available to a system. For example, morphogen-based SPH control takes advantage, in two key ways, of the behavior of morphogens in the aqueous environments typical of a growing embryo. First, it offloads pairwise distance estimates to the natural diminution with distance of a dilute substance as it diffuses and degrades. Second, it offloads a summation loop across many neighbors to the natural summation of dilute substances from multiple sources.

6.6 Morphogen Smoothing Functions

This section explores the use of morphogens to implement NSFs as a basis for SPH control. We refer to such NSFs as MSFs, as mentioned in

Sec. 6.3. We first discuss the properties of these MSFs with respect to SPH generally. Afterwards we derive the morphogen production rates needed to implement SPH control using MSFs.

Throughout this section, we make several simplifying assumptions:

- Each agent is a point.
- Each morphogen obeys Fick's law for homogeneous, isotropic diffusion (i.e., change in concentration is a constant proportion of the Laplacian).
- Degradation of each morphogen is uniform in the medium and proportional to morphogen concentration.
- Each agent is unmoving.
- Morphogen production, diffusion, and degradation are in equilibrium.

Later, we introduce corrections for some violations of the latter two assumptions.

For our embodied SPH approach to work, we need the SPH estimate of a field variable at a point to equal the summed concentration of an associated morphogen at that point:

$$\langle f(x) \rangle = \sum_j \frac{m_j}{\rho_j} f(x_j) W_j(x) = \sum_j c_j(x) \tag{15}$$

where $c_j(x)$ is the morphogen concentration at x attributable to neighboring agent j. Each agent must therefore choose to produce this morphogen at a rate r_j such that

$$c_j(x) = \frac{m_j}{\rho_j} f(x_j) W_j(x). \tag{16}$$

Let C_j be the total amount of morphogen in the environment due to agent j, at equilibrium:

$$C_j = \int_\Omega c_j(x) dx.$$

If k is the morphogen's decay rate, then C_j is related to r_j by

$$\frac{dC_j}{dt} = r_j - kC_j \tag{17}$$

and therefore reaches equilibrium when

$$C_j = \frac{r_j}{k} \tag{18}$$

Solving Eq. 16 for $W_j(x)$, integrating both sides, and considering the requirement of unity for the integral of $W_j(x)$, we have

$$\int_\Omega W_j(x)dx = \int_\Omega \frac{\rho_j}{m_j f(x_j)}c_j(x)dx = \frac{\rho_j}{m_j f(x_j)}C_j = \frac{r_j \rho_j}{km_j f(x_j)} = 1.$$

We can now solve for r_j to determine the correct production rate for agent j:

$$r_j = \frac{km_j f(x_j)}{\rho_j} \tag{19}$$

Under these simplified assumptions, the SPH smoothing functions implied by this production rate are described in Table 1.

Table 1. Idealized morphogen smoothing functions. The row labeled "Physical, normalized" refers to the physical part of the general solution to the relevant PDE, normalized so that its integral over space is unity. r is the magnitude of displacement. J_0 and Y_0 are Bessel functions of the first and second kind respectively, of order zero. K_0 is a modified Bessel function of the second kind, of order zero. E is diffusion rate and k is decay rate, and in the final row we make the substitution $h = \sqrt{\dfrac{E}{k}}$ to obtain forms resembling the typical presentation of SPH smoothing functions

	One dimension	Two dimensions	Three dimensions
Radial form of PDE	$E\dfrac{\partial^2 \rho}{\partial r^2} - k\rho = 0$	$\dfrac{E}{r}\dfrac{\partial \rho}{\partial r} + E\dfrac{\partial^2 \rho}{\partial r^2} - k\rho = 0$	$2\dfrac{E}{r}\dfrac{\partial \rho}{\partial r} + E\dfrac{\partial^2 \rho}{\partial r^2} - k\rho = 0$
General solution	$\rho(r) = c_1 e^{-\sqrt{\frac{k}{E}}r} + c_2 e^{\sqrt{\frac{k}{E}}r}$	$\rho(r) = c_1 J_0\left(ir\sqrt{\dfrac{k}{E}}\right) + c_2 Y_0\left(-ir\sqrt{\dfrac{k}{E}}\right)$	$\rho(r) = c_1 \dfrac{e^{-\sqrt{\frac{k}{E}}r}}{r} + c_2 \dfrac{e^{\sqrt{\frac{k}{E}}r}}{\sqrt{kr}}$
Physical, normalized	$\rho(r) = \dfrac{1}{2}\sqrt{\dfrac{k}{E}}e^{-\sqrt{\frac{k}{E}}r}$	$\rho(r) = \dfrac{k}{2\pi E}K_0\left(r\sqrt{\dfrac{k}{E}}\right)$	$\rho(r) = \dfrac{k}{4\pi E}\dfrac{e^{-\sqrt{\frac{k}{E}}r}}{r}$
With $h = \sqrt{\dfrac{E}{k}}$	$\rho(r) = \dfrac{e^{-\frac{r}{h}}}{2h}$	$\rho(r) = \dfrac{K_0\left(\dfrac{r}{h}\right)}{2\pi h^2}$	$\rho(r) = \dfrac{e^{-\frac{r}{h}}}{4\pi h^2 r}$

6.7 Correcting for Transient Spurious Morphogen Smoothing Functions

Traditional SPH assumes no lag time in information flow, but two factors prevent morphogens from conveying information instantaneously among agents. The first is the time needed for production and degradation processes to reach steady state, which is a non-spatial consideration. The second is the time needed for diffusion to spread information spatially from an agent to its neighbors. These two factors lead to transient spurious MSFs, and we discuss each in turn. In neither case do we fully correct for transient effects; instead, we make simplifying assumptions that allow us to correct for the most detrimental aspect of each effect.

6.7.1 Non-equilibrium of production and degradation rates

We first address the non-spatial transient effect of morphogen production being out of equilibrium with degradation, which occurs whenever production rates have recently changed. We make two simplifying assumptions to find a correction term for agents to add to their sensor readings to better approximate what those readings would be under equilibrium conditions. First, we neglect diffusion time; in other words, we assume that each agent instantaneously injects morphogen into its entire neighborhood according to the distribution of the idealized MSF. Second, we assume that the field variable values for all agents in a neighborhood are changing at the same rate a so that $f(x_j, t) = at + b_j$ since time $t = 0$, and that until $t = 0$ the values were not changing and the system was in equilibrium.

Based on these assumptions, we wish to find a correction term $g(t)$ such that $\langle f(x, t) \rangle = \Sigma_j c_j(x, t) + g(t)$. At a given time $t > 0$, from Eq. 17 and Eq. 19 the total mass of morphogen in the environment due to agent j obeys

$$\frac{dC_j(t)}{dt} = r_j(t) - kC_j(t) = \frac{km_j(at + b_j)}{\rho_j} - kC_j(t)$$

This equation is solved by

$$C_j(t) = \alpha e^{-kt} + \frac{m_j}{\rho_j}\left(at + b_j - \frac{a}{k}\right) \tag{20}$$

Because the system was in equilibrium at $t = 0$, we can substitute from Eq. 18 and Eq. 19:

$$C_j(0) = \frac{r_j(0)}{k} = \frac{m_j f(x_j, 0)}{\rho_j} = \frac{m_j b_j}{\rho_j}$$

After substituting into Eq. 20 and solving for α, we have:

$$C_j(t) = \frac{m_j a}{\rho_j k}e^{-kt} + \frac{m_j}{\rho_j}\left(at + b_j - \frac{a}{k}\right) = \frac{m_j a}{\rho_j k}e^{-kt} + \frac{m_j f(x_j, t)}{\rho_j} - \frac{m_j a}{\rho_j k},$$

which we can solve for $f(x_j, t)$:

$$f(x_j, t) = \frac{\rho_j}{m_j} C_j(t) + \frac{a}{k}(1 - e^{-kt}) \tag{21}$$

Substituing Eq. 21 into Eq. 12 and rearranging, a correct SPH estimate of $f(x, t)$ under our non-equilibrium assumptions would therefore be

$$\langle f(x, t) \rangle = \sum_j C_j(t) W_j(x) + \sum_j \frac{m_j}{\rho_j} \frac{a}{k}(1 - e^{-kt}) W_j(x) \tag{22}$$

The left summation of Eq. 22 is the SPH estimate of $f(x, t)$ under the equilibrium assumption and is also $\sum_j c_j(x, t)$, the sensed morphogen concentration at x (because we derived our production rate under the equilibrium assumption). The right summation of Eq. 22 is just the SPH estimate of $\frac{a}{k}(1 - e^{-kt})$, but this does not depend on j, so no SPH estimate is necessary for this term:

$$\langle f(x, t) \rangle = \sum_j C_j(t) W_j(x) + \frac{a}{k}(1 - e^{-kt}).$$

Thus we have our desired correction term:

$$g(t) = \frac{a}{k}(1 - e^{-kt}). \tag{23}$$

We wish for each agent to keep track of $g(t)$, which is possible so long as it maintains an estimate of $\frac{\partial f(x, t)}{\partial t}$. (In our simulation we simply have each agent remember one previous value of $f(x, t)$ which it can divide by a known timestep.) To see this, we first take a time derivative: $\frac{dg(t)}{dt} = ae^{-kt}$. Note that we can rearrange Eq. 23 to get $e^{-kt} = 1 - \frac{k}{a}g(t)$. We substitute this into the derivative along with $a = \frac{\partial f(x, t)}{\partial t}$ to obtain

$$\frac{dg(t)}{dt} = \frac{\partial f(x, t)}{\partial t} - kg(t) \tag{24}$$

which implies the needed update rule for agents to follow (in our simulation, $\Delta g(t) = \Delta f(x, t) - kg(t)\Delta t$).

6.7.2 Transient effects of dffusion rate

The time it takes for morphogen diffusion to convey information spatially also leads to violations of steady-state assumptions whenever the position or production rate of an agent has recently changed. Here we address only transient effects due to agent motion, because in our simulations it appears to be more detrimental than spatial effects due to production rate changes. As an agent moves, its contribution to the total morphogen field becomes stretched behind and compressed in front, with more complex distortions if the agent's path curves. Even more complex effects arise as a result of the particularities of an agent's instrument positions and the way a given medium flows around an agent's body. Because such particularities imply implementation-dependence of distortions, we do not pursue analytic corrections. Instead, we propose that agents be calibrated to correct for some of this distortion, an approach we test in simulation.

The most damaging aspect of this distortion is the effect that an agent's own contribution to the morphogen field has on its gradient estimates. In addition, an agent has no knowledge of its neighbors' individual positions and velocities, limiting its ability to correct for distortions in neighbors' contributions to the morphogen field. For these reasons we focus calibration on correcting for distortions in an agent's own contributions to its gradient estimates.

The central idea of calibration is to have isolated agents move at constant velocities and, once steady-state relative to the agent is reached, to record the measured gradient. A table of gradients for different speeds can be created in this way, and during operation these gradients (with appropriate interpolation) can be subtracted from measured gradients to partially correct for motion-related asymmetry in the MSF. For our simulation, we repeat this procedure with different headings to average out numerical effects related to the finite difference method grid used for diffusion.

6.8 Beyond Fluid Dynamics: SPH as a Partial Morphogen Compiler

SPH robotic control generally, and our NSF variant in particular, can be viewed as a compilation technique for Morphogen code. The utility of SPH as a compilation technique is limited in the present work by the requirement that swarm density fields must obey the mass continuity equation. $\frac{\partial \rho}{\partial t} = -\nabla \cdot (\rho u)$. This trivially holds when swarm density is specified in Morphogen implicitly by velocity or acceleration fields, but may not hold when swarm density fields are specified directly. For example, a nonconservative equation such as $\frac{\partial \rho}{\partial t} = 5$ cannot yet be compiled with our method. Other fields representing swarm state are not limited in this way. In future work we plan to relax this requirement through agent recruitment and removal or through cell-like division and apoptosis.

SPH compilation introduces a layer of abstraction or virtualization between what we might think of as Morphogen "application" code and what we envision as a library of physical morphogens, which can also be described in Morphogen. Consider a Morphogen program that specifies a velocity or acceleration field for a swarm, along with several other fields. These other fields may be thought of as representing tissue differentiation, gene expression, or even morphogens, as seen in earlier examples. To compile such a program into agent code, all non-swarm fields are treated as agent state, and a unique physical "library" morphogen is associated with each. (In practice, a programmer might find it convenient to specify these associations, because physical morphogens with different ratios of diffusion to degradation rates might prove more robust for implementing different fields. However, such low-level associations are not essential in principle: Eq. 19 specifies the correct production rate regardless of diffusion and degradation rates.) As described in Sec. 6.6, these physical morphogens implicitly provide the smoothing functions needed for each agent to make SPH estimates of a given field and its derivatives.

To further illustrate this layer of abstraction, consider a Morphogen application program that specifies what the programmer conceptualizes as a morphogen (for example, if the Morphogen code specifies a diffusion and degradation rate as well as sources and sinks). The SPH compilation process would nonetheless treat this field as a swarm state variable and associate it with a physical morphogen, whose physical parameters are theoretically independent of the abstract morphogen described by the application code. This layer of abstraction frees the programmer from the need to find a physical substance with particular parameters. Instead, we envision engineering a small library of physical morphogens that can be repurposed to implement a variety of fields in different Morphogen programs.

As mentioned in Sec. 6.2, SPH swarm robotic control has focused on enabling swarms to behave somewhat like fluids, as described by the Navier-Stokes or Euler equations. Swarms have therefore been specified with acceleration fields, and other fields have been designed to simulate such physical quantities as energy and pressure. In the context of our current work, we prefer to reframe these applications as special cases of SPH compilation of implicit Morphogen-like programs that happen to describe roughly physical fluid-like behavior. (This is a simplification; for example, Pimenta et al. [39] derive one force using SPH, and add to it a body force derived separately from a potential field.)

In comparison to past fluid-related SPH swarm control, the example that follows illustrates some of this potential generality of SPH compilation. Rather than specifying physics-inspired quantities such as energy or pressure, arbitrary fields useful in solving a path-finding problem are specified. And rather than specifying acceleration, velocity is instead specified, making swarm behavior more similar to flocking than to Newtonian dynamics. (Velocity-based motion is also a natural fit for fluid media with low Reynolds numbers such that propulsion force is approximately proportional to velocity rather than to acceleration. However, agents could integrate or differentiate on their own to achieve, respectively, acceleration-based control where force is proportional to velocity or velocity-based control where force is proportional to acceleration.)

6.9 Example: Path-finding

To illustrate our method, we simulate a two-dimensional environment that is inspired by, but does not rigorously implement, an aqueous environment with low Reynolds number, as would be found in a developing embryo. Agents are modeled as discs. Sensors are placed around the surface at 90° intervals, with morphogen production instruments on the surface halfway between each pair of sensors. Agents have full control over their velocity, but no propulsion mechanism is explicitly modeled; agents' locations are simply changed. As agents move, the surrounding medium and morphogens flow around the agents' bodies. This flow is qualitatively similar to laminar flow but does not strictly obey fluid dynamics. Motion of the aqueous medium and morphogens is calculated using a finite difference method. Some random noise is applied to agents' motion to qualitatively simulate Brownian motion, and some bias and noise is applied to sensor readings.

    In the Morphogen program below, $\langle T \rangle$ refers to the SPH estimate of T, as opposed to an agent's internal stored value.

substance swarm :
    scalar fields :
          ρ // swarm density
          A // environmental cue for lower-left square
          B // environmental cue for upper-right square
          S // to diffuse from lower-left square
          T // to follow gradient of S back from upper-right square
    vector fields :
          $\mathbf{V}$ // swarm velocity

    behavior :
    params :
          $\theta_\rho = 0.02$
          $\theta_A = 0.2$
          $\theta_B = 0.2$
          $\theta_{T_1} = 0.1$
          $\theta_{T_2} = 0.05$
          $\kappa_{S_1} = 500$
          $\kappa_{S_2} = 500$
          $\kappa_{S_3} = 1$
          $\kappa_{S_4} = 100$

$$\kappa_{T_1} = 500$$
$$\kappa_{T_2} = 600$$
$$\kappa_{T_3} = 0.002$$
$$\kappa_{T_4} = 60$$
$$\kappa_V = 0.0003$$
$$\kappa_{V_{\max}} = 0.003$$

$$\textbf{let } \phi = \arccos\left(\frac{\nabla S \cdot \nabla T}{\|\nabla S\| \|\nabla T\|}\right)$$

$$\mathrm{D}S\,-\,= [\rho < \theta_\rho \vee B > \theta_B]\kappa_{S_1} S$$
$$\mathrm{D}S\,+\,= [A > \theta_A \vee S > 1]\kappa_{S_2}(1 - S)$$
$$\mathrm{D}S\,+\,= [A \le \theta_A](\kappa_{S_3}\nabla^2 S - \kappa_{S_4}S)$$
$$\mathrm{D}T\,+\,= [B > \theta_B]\kappa_{T_1}(1 - T)$$

$$\mathrm{D}T\,+\,= [\|\nabla S\| > 0 \wedge \|\nabla T\| > 0 \wedge \langle T\rangle > \theta_{T_1}]\left(\frac{\phi}{\pi}\right)^6\kappa_{T_2}$$
$$\mathrm{D}T\,+\,= \kappa_{T_3}\nabla^2 T$$
$$\mathrm{D}T\,-\,= \kappa_{T_4}T^2$$
$$\mathrm{D}T\,+\,= (\mathbf{V}\cdot\nabla T)[\langle T\rangle > \theta_{T_2} \wedge T < \theta_{T_1}]$$

$$\textbf{let } \mathbf{V}_0 = \frac{1}{\rho}\kappa_V\,2T\ln\left(\frac{8\rho}{T}\right)\left(\frac{\nabla\rho}{T} - \frac{\rho\nabla T}{T^2}\right)$$

$$\mathbf{V} = \kappa_{V_{\max}}\tanh\left(\frac{\|\mathbf{V}_0\|}{\kappa_{V_{\max}}}\right)\frac{\mathbf{V}_0}{\|\mathbf{V}_0\|}$$

Simulation of the above Morphogen program is shown in Table 2. All PDEs and simulation parameters, including starting density, are the same for all three runs shown, with only agent diameter and number varying. To hold starting density roughly constant across the eightfold difference in agent diameter, agent number varies about 64-fold. The qualitative similarity across this range demonstrates the potential of Morphogen and SPH to facilitate scalable swarm dynamics.

We describe some of the key components of this program, which aligns agents along the shortest path between two distinct environmental cues. Agents that sense a cue A represented by the lower-left square set a high value for the variable S. Other agents implement a diffusion-decay equation to establish the field distribution of S. (This equation is independent of the diffusion-decay equation that models the physical behavior of the morphogen that implements S in the SPH scheme.) Agents that sense a second cue B represented by the upper-right square as well as a non-zero gradient of S set a high value for variable T. Other agents increase their value of T depending on how nearly opposite are the gradient directions of S and T and whether a neighborhood estimate

Table 2. Path-finding example. Agents use SPH to establish path between upper right and lower left squares

	166 agents, diameter 0.016	*1186 agents, diameter 0.006*	*10675 agents, diameter 0.002*
Time = 0			
Time = 20			
Time = 70			
Time = 270			

of T is above a threshold. Otherwise, T diffuses slowly. Agents follow the gradient of a function involving both T and local density such that a target density is achieved where T is high. Agents' speed saturates with the tanh function so that dynamics are similar to flocking behavior when far from optimum arrangement but approach rest when closer to that optimum.

Because of the Lagrangian nature of SPH, field variables are naturally carried along with agents, which may or may not be desired. At low speeds and with field dynamics that are somehow (as in this program) tethered to environmental cues, the distinction is often subtle in practice. When it is desired that Morphogen PDEs refer to an Eulerian frame, advection can be added counter to agent velocity, as reflected by $ÐT \mathrel{+}= \mathbf{V} \cdot \nabla T$ in the program above. In this program, we only add this counter-advection where it helps the growing T region to continue its growth against the opposing motion of agents following the T gradient.

7. Conclusions

We have argued that embryological morphogenesis provides a model of how massive swarms of microscopic agents can be coordinated to assemble complex, multiscale hierarchical structures, that is, artificial morphogenesis or morphogenetic engineering. This is accomplished by understanding natural morphogenetic processes in mathematical terms, abstracting from the biological specifics, and implementing these mathematical principles in artificial systems.

As have embryologists, we have found partial differential equations and continuum mechanics to be powerful mathematical tools for describing the behavior of very large numbers of very small agents, in fact, taking them to the continuum limit. In this way we intend to have algorithms that scale to very large swarms of microrobots.

To this end we have developed a PDE-based notation for artificial morphogenesis and designed a prototype morphogenetic programming language. This language permits the precise description of morphogenetic algorithms and their automatic translation to simulation software, so that morphogenetic processes can be investigated.

We illustrated the morphogenetic programming language and morphogenetic programming techniques with two examples. The first addressed the problem of routing dense bundles of many fibers between specified regions of an artificial brain. Inspired by axonal routing during embryological development, we used a modified flocking algorithm to route fiber bundles between origins and destinations while avoiding other bundles. Simulations showed that this algorithm scaled over at least four orders of magnitude, with swarms of 5000 agents. Then we took the number and size of agents to the continuum limit and showed how morphogenetic programming could be used to coordinate a massive swarm of agents to lay down a path between designated termini while avoiding obstacles.

Our second example showed how a natural morphogenetic model—the clock-and-wavefront model of spinal segmentation—could be applied in morphogenetic engineering both in a similar context—creating the segmented "spine" of an insect-like robot body—and for a different purpose—assembling segmented legs on the robot's spine. A massive swarm of microscopic agents is supplied from external sources and guided to the assembly sites, where they begin a process of differentiation coordinated by the emission of and response to several morphogens. As demonstrated in simulation, this reasonably complex process can be controlled to assemble a structure with a spine and legs with specified numbers and sizes of segments.

Finally, we showed how an embodied variation of smoothed particle hydrodynamics (SPH) swarm robotic control can be applied to the global-to-local compilation problem, that is, the derivation of individual agent control from global PDE specifications. By these means, available physical morphogens can be used to implement the abstract morphogens and other fields required for a morphogenetic process. The physical morphogens define natural smoothing functions that can be used in SPH to estimate the concentrations, gradients, and other functions of morphogenetic fields.

References

1. H. Abelson, D. Allen, D. Coore, C. Hanson, G. Homsy, T.F. Knight Jr., R. Nagpal, E. Rauch, G.J. Sussman and R. Weiss. Amorphous Computing. Commun. ACM 43.5 (May 2000), 74–82 (2000).
2. P. Bourgine and A. Lesne (eds). Morphogenesis: Origins of Patterns and Shapes. Berlin: Springer (2011).
3. J. Cooke and E.C. Zeeman. A clock and wavefront model for control of the number of repeated structures during animal morphogenesis. J. Theoretical Biology 58, 455–476 (1976).
4. P.G. de Gennes. Soft Matter. Science 256, 495–497 (1992).
5. M.-L. Dequant and O. Pourqui. Segmental patterning of the vertebrate embryonic axis. Nature Reviews Genetics 9, 370–382 (2008).
6. A. Doostmohammadi, J. Igns-Mullol, J.M. Yeomans and F. Sagus. Active nematics. Nature Communications 9.1, 3246 (2018).
7. R. Doursat. Organically grown architectures: creating decentralized, autonomous systems by embryomorphic engineering. *In*: R.P. Wrtz (ed.), Organic Computing. pp. 167–200. Springer (2008).
8. R. Doursat, Hiroki Sayama and Olivier Michel. A review of morphogenetic engineering. Natural Computing 12.4 (Dec. 2013), 517–535 (2013).
9. K.W. Fleischer. A Multiple-mechanism Developmental Model for Defining Self-organizing Geometric Structures. PhD thesis. California Institute of Technology (1995).
10. G. Forgacs and S.A. Newman. Biological Physics of the Developing Embryo. UK: Cambridge University Press, Cambridge (2005).
11. R. Fujiwara, T. Kano and A. Ishiguro. Self-swarming robots that exploit hydrodynamical interaction. Advanced Robotics (Feb. 5, 2014), 1–7 (2014).
12. J.-L. Giavitto and A. Spicher. Computer morphogenesis. *In*: P. Bourgine and A. Lesne (eds), Morphogenesis: Origins of Patterns and Shapes, pp. 315–340. Berlin: Springer (2011).
13. R.A. Gingold and J.J. Monaghan. Smoothed particle hydrodynamics: theory and application to non-spherical stars. Monthly notices of the Royal Astronomical Society 181.3, 375–389 (1977).
14. S.C. Goldstein, J.D. Campbell and T.C. Mowry. Programmable matter. Computer 38.6 (June 2005), 99–101 (2005).

15. H. Hamann. Space-time continuous models of swarm robotic systems. *In*: Rdiger Dillmann, David Vernon, Yoshihiko Nakamura and Stefan Schaal (eds), Cognitive Systems Monographs. Vol. 9. Berlin, Heidelberg: Springer Berlin Heidelberg (2010).

16. C. Huang, J.M. Lei, M.B. Liu and X.Y. Peng. An improved KGF-SPH with a novel discrete scheme of Laplacian operator for viscous incompressible fluid flows. International Journal for Numerical Methods in Fluids 81.6 (June 30, 2016), 377–396 (2016).

17. H. Kitano. Morphogenesis for evolvable systems. *In*: E. Sanchez and M. Tomassini (eds), Towards Evolvable Hardware: The Evolutionary Engineering Approach, pp. 99–117. Berlin: Springer (1996).

18. J.W. Lichtman and W. Denk. The big and the small: challenges of imaging the brain's circuits. Science 334.6056 (Apr. 2011), 618–623 (2011).

19. G.R. Liu and M.B. Liu. Smoothed particle hydrodynamics: a mesh-free particle method. OCLC: ocm52947194. 449 pp. New Jersey: World Scientific (2003).

20. L.B. Lucy. A numerical approach to the testing of the fission hypothesis. The Astronomical Journal 82, 1013–1024 (1977).

21. B.J. MacLennan. A morphogenetic program for path formation by continuous flocking. International Journal of Unconventional Computing 14, 91–119 (2019).

22. B.J. MacLennan. Artificial morphogenesis as an example of embodied computation. International Journal of Unconventional Computing. 7.1–2, 3–23 (2011).

23. B.J. MacLennan. Coordinating Massive Robot Swarms. International Journal of Robotics Applications and Technologies 2.2, 1–19 (2014).

24. B.J. MacLennan. Coordinating swarms of microscopic agents to assemble complex structures. *In*: Ying Tan (ed.), Swarm Intelligence, Vol. 1: Principles, Current Algorithms and Methods. PBCE 119. Institution of Engineering and Technology, Chap. 20, pp. 583–612 (2018).

25. B.J. MacLennan. Embodied computation: applying the physics of computation to artificial morphogenesis. Parallel Processing Letters 22.3, pp. 1240013 (2012).

26. B.J. MacLennan. Models and mechanisms for artificial morphogenesis. *In*: F. Peper, H. Umeo, N. Matsui and T. Isokawa (ed.), Natural Computing. Springer series, Proceedings in Information and Communications Technology (PICT) 2. Tokyo: Springer (2010).

27. B.J. MacLennan. Molecular coordination of hierarchical self-assembly. Nano Communication Networks 3.2 (June 2012), 116–128 (2012).

28. B.J. MacLennan. Morphogenesis as a model for nano communication. Nano Communication Networks 1.3, pp. 199–208 (2010).

29. B.J. MacLennan. Preliminary development of a formalism for embodied computation and morphogenesis. Technical Report UT-CS-09-644. Knoxville, TN: Department of Electrical Engineering and Computer Science, University of Tennessee (2009).

30. B.J. MacLennan. The morphogenetic path to programmable matter. Proceedings of the IEEE 103.7, 1226–1232 (2015).

31. B.J. MacLennan. The Synmac Syntax Macroprocessor: Introduction and Manual, Version 5. Faculty Publications and Other Works—EECS. http://

trace.tennessee.edu/utkelecpubs/23. University of Tennessee, Department of Electrical Engineering and Computer Science (2018).

32. J.M.Z. Maningo, G.E.U. Faelden, R.C.S. Nakano, A.A. Bandala, R. Rhay P. Vicerra and E.P. Dadios. Formation control in quadrotor swarm aggregation using smoothed particle hydrodynamics. Region 10 Conference (TENCON), 2016 IEEE, 2070–2075 (2016).

33. S. Murata and H. Kurokawa. Self-reconfigurable robots: shape-changing cellular robots can exceed conventional robot flexibility. IEEE Robotics & Automation Magazine (Mar. 2007), 71–78 (2007).

34. R. Nagpal, A. Kondacs and C. Chang. Programming methodology for biologically-inspired self-assembling systems. AAAI Spring Symposium on Computational Synthesis: From Basic Building Blocks to High Level Functionality (Mar. 2003).

35. D. Needleman and Z. Dogic. Active matter at the interface between materials science and cell biology. Nature Reviews Materials 2 (July 2017), 17048 EP (2017).

36. H. Oh, A.R. Shirazi, C. Sun and Y. Jin. Bio-inspired self-organising multi-robot pattern formation: a review. Robotics and Autonomous Systems 91 (May 2017), 83–100 (2017).

37. M.R. Pac, A.M. Erkmen and I. Erkmen. Control of robotic swarm behaviors based on smoothed particle hydrodynamics. Intelligent Robots and Systems, 2007. IROS 2007. IEEE/RSJ International Conference on IEEE, pp. 4194–4200 (2007).

38. Ja. R. Perkinson and B. Shafai. A decentralized control algorithm for scalable robotic swarms based on mesh-free particle hydrodynamics. Proc. of the IASTED Int. Conf. on Robot and Applications. pp. 102–107 (2005).

39. L.C.A. Pimenta, G.A.S. Pereira, N. Michael, R.C. Mesquita, M.M. Bosque, L. Chaimowicz, and V. Kumar. Swarm coordination based on smoothed particle hydrodynamics technique. IEEE Transactions on Robotics 29.2 (Apr. 2013), 383–399 (2013).

40. I. Salazar-Ciudad, J. Jernvall and S.A. Newman. Mechanisms of pattern formation in development and evolution. Development 130, 2027–2037 (2003).

41. M.B. Silic, Z. Song and K. Mohseni. Anisotropic flocking control of distributed multi-agent systems using fluid abstraction. 2018 AIAA Information Systems-AIAA Infotech @ Aerospace. American Institute of Aeronautics and Astronautics (2018).

42. A. Spicher, O. Michel and J.-L. Giavitto. Algorithmic self-assembly by accretion and by carving in MGS. Proc. of the 7th International Conference on Artificial Evolution (EA '05). Lecture Notes in Computer Science 3871, pp. 189–200. Berlin: Springer-Verlag (2005).

43. L.A. Taber. Nonlinear Theory of Elasticity: Applications in Biomechanics. Singapore: World Scientific (2004).

44. B.P. Teague, Patrick Guye and Ron Weiss. Synthetic morphogenesis. Cold Spring Harbor Perspectives in Biology (June 7, 2016), a023929 (2016).

45. U. Tilki and A.M. Erkmen. Fluid swarm formation control for hand gesture imitation by ellipse fitting. Electronics Letters 51.6, 473–475 (2015).

46. Wang Xi, Thuan Beng Saw, Delphine Delacour, Chwee Teck Lim and Benoit Ladoux. Material approaches to active tissue mechanics. Nature Reviews Materials 4.1, 23–44 (2019).

3

Ant Cemeteries as a Cluster or as an Aggregate Pile

Tomoko Sakiyama

Department of Information Systems Science, Faculty of Science and Engineering, Soka University, Tokyo 192-8577, Japan

Email: sakiyama@soka.ac.jp

1. Background

Necrophoresis is a behavior found in social insects. They carry a corpse, i.e., the dead bodies of members of their colony from the nest to prevent infection from spreading throughout the colony [1, 2]. Necrophoresis by ant workers can lead to the formation of cemeteries. This phenomenon results from simple clustering rules. If ant workers find a corpse they then pick-up that item with a probability that will decrease with cluster size; while corpse-carrying ants drop their carrying corpses with a probability that will increase with cluster size. These local amplification processes result in the formation of large piles of corpses [3].

Previous studies have assumed a complex function regarding the time evolution of a local cluster of corpses and agents have to strictly estimate local cluster size to determine the drop rate [3]. Even if that system looks like another system at first glance and at the macro-level, it can be important to examine how to model the decision-making processes of the agents in the system [4]. In fact, several aggregate behaviors of foraging ants seem to have been revised due to the mismatch between observed phenomena at the agent-level and nonlinearity of classical models [4, 5].

Recently, I developed an agent-based model which illustrated the pile growth of ants using simpler rules [6]. In that study, I developed one-dimensional pile growth models of artificial ants and introduced decision-making by agents in an easy manner. These artificial ants detected a certain number of corpses around them and determined to drop a carried corpse based on that value. In addition, the artificial ants coordinated the probability of drop, based on the detection of their nest-mates, resulting in the growth of clusters of agents.

Here, I focus on the spatial distribution of clusters in my proposed model and demonstrate that cluster size follows a distinctive distribution.

2. Methods

A description of the model follows [6].

Artificial agents that move in a one-dimensional discrete lattice were simulated. As boundary conditions, periodic boundaries were assumed. I randomly introduced N_{alive} agents and N_{dead} corpses on the field. Here, I explain sub-models. In each sub-model, agents update their positions synchronously.

Corpse detection

The agent k can detect a corpse only if that corpse locates the current position of the agent k.

if a corpse locates on x_t^k,

then $detect\_c_t^k \rightarrow 1$ and calculates $sum\_c_t^k$, $\qquad\qquad$ (1)

or else $detect\_c_t^k \rightarrow 0$. $\qquad\qquad\qquad\qquad\qquad\qquad\qquad\qquad$ (2)

Here, x_t^k represents the current position of the agent k at time t while $detect\_c_t^k$ determines whether or not the agent k detects a corpse at time t. The parameter $sum\_c_t^k$ indicates the total number of corpses detected by the agent k at time t.

Nest-mates detection

The agent k can detect a nest-mate only if that nest-mate locates on the current position of agent k. At this time, the following equations are satisfied.

if a nest-mate locates on x_t^k,

$$\text{then } detect\_n_t^k \rightarrow 1, \qquad\qquad\qquad (3)$$

or else $\qquad\qquad\qquad detect\_n_t^k \rightarrow 0. \qquad\qquad\qquad (4)$

Here, $detect\_n_t^k$ determines whether or not the agent k detects a nest-mate at time t.

Probability calculation

The probability of corpse-dropping/picking up for the agent k is determined in the following manner:

if $carry_t^k = 1$ AND $detect\_c_t^k = 1$,

$\qquad\qquad$ if $detect\_n_t^k = 1$,

if $sum\_c_t^k \neq 0$ AND $sum\_c_t^k < threshold$,

$$\text{then } prob_t^k = P_{\text{low}}, \tag{5}$$

else if $sum\_c_t^k >= threshold$,

$$\text{then } prob_t^k = P_{\text{high}}. \tag{6}$$

else if $detect\_n_t^k = 0$,
if $sum\_c_t^k \neq 0$ AND $sum\_c_t^k < threshold$,

$$\text{then } prob_t^k = P_{\text{high}}, \tag{7}$$

else if $sum\_c_t^k >= threshold$,

$$\text{then } prob_t^k = P_{\text{low}}. \tag{8}$$

else if $carry_t^k = 0$ AND $detect\_c_t^k = 1$,

$$\text{then } prob_t^k = P_{\text{pick}}, \tag{9}$$

Here, $carry_t^k$ determines whether the agent k caries a corpse or not, at time t while $prob_t^k$ indicates the probability of drop or pick up for the agent k at time t. The agent k detecting a corpse determines the probability of drop (P_{high} or P_{low}) based on the relationship between $sum\_c_t^k$ and $threshold$ if the agent caries a corpse ($carry_t^k = 1$). On the contrary, the agent k detecting a corpse determines the probability of pick up (P_{pick}) if the agent does not carry a corpse ($carry_t^k = 0$).

Moreover, according to the equations (5)–(8), the agent k replaces $P_{\text{high}}/P_{\text{low}}$ with $P_{\text{low}}/P_{\text{high}}$, which is dependent on the value of $detect\_n_t^k$. Please note that one agent is randomly chosen to pick up a corpse if two or more agents are going to pick up a corpse.

Picking up/Dropping

if $detect\_c_t^k = 1$,

if $carry_t^k = 1$ AND $rn_t^k <= prob_t^k$,

$$\text{the agent } k \text{ drops the carrying corpse and } carry_t^k \rightarrow 0, \tag{10}$$

else if $carry_t^k = 0$ AND $rn_t^k <= prob_t^k$,

$$\text{the agent } k \text{ picks up a detected corpse and } carry_t^k \rightarrow 1, \tag{11}$$

else if $detect\_c_t^k = 0$,

if $carry_t^k = 1$ AND $rn_t^k <= P_{\text{drop}}$,

$$\text{the agent } k \text{ drops the carrying corpse and } carry_t^k \rightarrow 0, \tag{12}$$

Here, rn_t^k indicates the random number satisfying the following:

$$rn_t^k \in [0.0, 1.0].$$

P_{drop} presents the probability of dropping a corpse when agents do not detect any corpses.

The equation (10)/(11) shows that the agent k detecting a corpse drops/picks up a corpse based on the calculated probability $prob_t^k$. The equation (12) shows that the agent k carrying a corpse drops the item with a constant probability P_{drop} even if the agent detects no corpses.

Position updating

The agent k updates the current position using following equation:

$$x_t^k \rightarrow x_t^k + 1 \text{ or } x_t^k - 1 \text{ with equal probability.} \tag{13}$$

Time updating

$$t \rightarrow t+1, \text{ for all agents.} \tag{14}$$

This proposed model is named as the switching model. It's Model description is the following:

STEP 1: Corpse detection
STEP 2: Nest-mates detection
STEP 3: Probability calculation
STEP 4: Picking up/Dropping
STEP 5: Position updating
STEP 6: Time updating and Go to STEP 1

3. Results

Parameters are shown in Table 1. Figure 1 which represents the evolution as a function of time of the mean number of piles in the switching model. Here I defined a cluster as a pile if it consisted of more than five corpses. Agents in the proposed models were allowed to detect only one lattice at each time. Therefore, piles on different lattices were dealt with separately. In the early stages, many piles are constructed and these disappear as time evolves.

Figure 2 represents an example of pile distribution according to x-axis. As shown in this figure, some dominant piles are constructed while smaller clusters remain (see Fig. 3 where all clusters including clusters whose size are less than 5 are shown).

To see the frequency distribution of clusters, I investigated the frequency distribution of cluster size. According to Fig. 4, which presents the relationship between cluster size and its cumulative frequency, cluster size appears to obey a power-law. To this end, smaller clusters remain and might enable the system to not enter a steady state.

Table 1. Parameters used for the analysis

Parameters	Value	Description
Φ	100	The number of agents
φ	300	The number of corpses
L	100	Field length
P_{high} / P_{low}	0.5 / 0.1	Probability of dropping a corpse when agents detect a corpse
P_{pick}	0.05	Probability of picking up a corpse
P_{drop}	0.05	Probability of dropping a corpse when agents do not detect any corpses
Threshold	15	Threshold value for choosing P_{high} or P_{low}
N of time step	10000	Number of time steps conducted for measurement
N of trials	100	Number of trials conducted for measurement

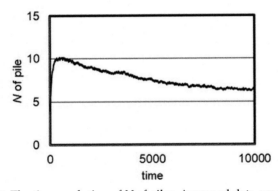

Fig. 1. The time evolution of N of piles. Averaged data are shown.

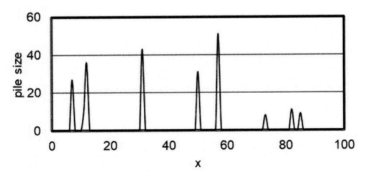

Fig. 2. An example of pile distribution according to x-axis.
Data from one trial is plotted.

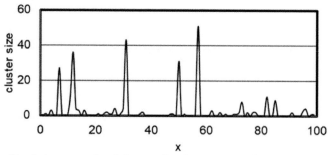

Fig. 3. An example of cluster distribution according to *x*-axis. Data from one trial is plotted.

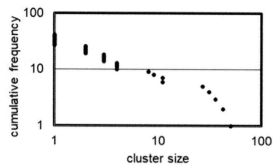

Fig. 4. An example of the relationship between cluster size and its cumulative frequency. Data from Fig. 2 and Fig. 3 is shown.

When agents are prohibited from replacing the P_{high}/P_{low} with P_{low}/P_{high}, the probability of drop can be fixed to a high value (≈ 0.50) due to P_{high} is set to 0.50. In fact, in such cases, actual probability of drop is approximately 0.50, which contradicts with results of ant experiments. Ants seem to drop a carrying corpse with lower probability; approximately 0.2~0.3. In my model, I observed that probability of drop when an agent was carrying a corpse was around 0.30. In that sense, modification of drop rate is essential to form a few dominant piles. Using a simpler model just like a threshold model does not necessarily produce a few piles with large size. This is perhaps because piles would grow in various locations if the probability of drop is fixed to a certain value. To this end, ants might make a decision using completely different information concurrently with a simple threshold rule. Such an awkward manner can bridge a gap between an individual behaviour and an aggregate behaviour.

4. Conclusion

I developed a pile growth model of artificial ants. In my proposed model, agents sometimes modify the probability of drop, which was then dependent on whether they detected or did not detect their nest-mates.

In previous studies, models assumed phenomenological results/functions obtained from experimental studies. However, recent studies regarding macro-behaviour appear to use a simple threshold rule rather than a complicated function. The important point to note is that agents change the background/assumption of rules. In that sense, agents are definitely not mechanical agents which blindly obey rules [4, 6]. My model might describe how to model the decision-making processes of agents in the pile-growth system of ants, which may bridge a gap between the simple individual behaviour at the micro-level and the complex patterns at the macro-level [4, 7].

My model represents the typical time evolution of piles. Thus, a few big size piles emerge as time goes on. On the other hand, I found that the cluster size followed a power-law distribution. In that sense, smaller size corpses also remain to some extent, which can contribute to reconstruction of the system without falling into a steady state.

References

1. L. Diez, P. Lejeune and C. Detrain. Keep the nest clean: survival advantages of corpse removal in ants. Biology Letters 10(7), 20140306–20140306 (2014).
2. C. Jost, J. Verret, E. Casellas, J. Gautrais, M. Challet, J. Lluc, S. Blanco, M.J. Clifton and G. Theraulaz. The interplay between a self-organized process and an environmental template: corpse clustering under the influence of air currents in ants. Journal of The Royal Society Interface 4(12), 107–116 (2007).
3. G. Theraulaz, E. Bonabeau, S.C. Nicolis, R.V. Solé, V. Fourcassié, S. Blanco, R. Fournier, J.L. Joly, P. Fernández, A. Grimal, P. Dalle and J.L. Deneubourg. Spatial patterns in ant colonies. Proc. Natl Acad. Sci. USA 99, 9645–9649 (2002).
4. T. Sakiyama. Ant droplet dynamics evolve via individual decision-making. Scientific Reports 7, 14877-1– 14877-8 (2017).
5. A. Perna et al. Individual rules for trail pattern formation in argentine ants (*Linepithema humile*). PLoS Comput Biol 8, e1002592, 1–12 (2012).
6. T. Sakiyama. Interactions between worker ants may influence the growth of ant cemeteries. Submitted.
7. E.J.H. Robinson, N.R. Franks, S. Ellis, S. Okuda and J.A.R. Marshall. A simple threshold rule is sufficient to explain sophisticated collective decision-making. PLOS ONE 6(5), e19981 (2011).

4

Robust Swarm of Soldier Crabs, *Mictyris guinotae*, Based on Mutual Anticipation

Y.-P. Gunji[1]*, H. Murakami[2], T. Niizato[3], Y. Nishiyama[4], K. Enomoto[5], A. Adamatzky[6], M. Toda[7], T. Moriyama[8] and T. Kawai[1]

[1] Department of Intermedia, Art and Science, Faculty of Fundamental Science and Engineering, Waseda University Ohkubo 3-4-1, Shinjuku-ku, Tokyo 657-8501, Japan

[2] Research Center for Advanced Science and Technology, The University of Tokyo Komaba 4-6-1, Meguro-ku, Tokyo, 153-0041, Japan

[3] Information and Systems, Faculty of Engineering, University of Tsukuba Tenoudai 1-1-1, Tsukuba Science City, 305-8573, Japan

[4] Information and Management Systems Engineering, Nagaoka University of Technology Kamitomioka-chou 1603-1, Nagaoka, 940-2188, Japan

[5] Department of Electronic Systems Engineering, School of Engineering, University of Shiga prefecture Yasaka-machi 2500, Hikone, 522-0057, Japan

[6] Center for Unconventional Computing, University of the West of England, Bristol, BS16 1QY, UK

[7] Center for Management of Information Technologies, Kumamoto University, Kurokami 2-39-1, Kumamoto City, Kumamoto, Japan

[8] Department of Bioengineering, Shinshu University, Ueda 386-8567, Japan

1. Introduction

Swarms, flocks and schools are some of the most intriguing collective behaviors [12, 50, 52, 54]. A flock of starlings behaves as if it had one body and/or one consciousness [11]. It has recently become possible to obtain kinetic data with regard to how real organisms move using image analysis techniques and high-resolution video cameras [1, 2, 9, 10]. The collective intelligence of animal groups has been addressed by modeling communal decision-making using computer models [13, 29, 39], including Boids [44] and self-propelled particles (SPP) [52, 53]. The synchronized taking

*Corresponding author: yukio@waseda.jp

off and landing of a flock [50] and its decisions about the homing route [13] can be explained using variants of these computer models. It may be easier to view the behavior of a flock or school as a self-organizing process that, under special condition patterns which extend well beyond the scale of the components, can arise spontaneously from interactions between the components.

Although self-organization is a powerful tool for understanding the complex systems underlying collective behavior, the difference between a physical component and a living individual in an animal group has recently been addressed [16]. This difference stems from the diversity of the individuals, which can contribute to the internal structure of an animal group. In a flock of birds or school of fish, individuals may be strongly correlated as whole. They may exhibit similar velocities without divergence, making it possible for a flock to move as though it had one body [44, 50, 52, 53, 54]. A flock, school or swarm thus appears to be homogeneous. However, this homogeneity may be the result of heterogeneity, which can result from the diversity of the individuals. For example, in homing pigeons, there is a hierarchy of flock-mates that correlates with the spatial position of the individuals [39]. A starling flock contains a scale-free sub-domain, in which flock-mates are positively correlated. In ant colonies, various forms of heterogeneity arise from the castes to which individuals belong, the experience of each worker and their genetic code, which can contribute to the parallel-distributed processing as whole [16]. In a starling flock, individuals can interact with their flock-mates, despite the varying distances between the flock-mates due to the topological distance [2].

Inherent noise results in the divergence of the individuals' actions, which can cooperate to allow the coherent motion of an animal group. Recently, it has been suggested that inherent turbulence may play an essential role in collective motion [4, 7, 49, 55]. In ant colonies, noise arises in obedience to trail pheromones, encouraging foraging [8]. An adequate balance between the recruitment to and the retreat from a trail can be optimized at a specific noise level. Thus, a flock, school or swarm can be a parallel-distributed system driven by inherent perturbation. In contrast, external noise, such as thermal noise, is assumed to influence individuals in a flock or school homogenously. This concept is integrated into computer models, which couple local interaction for alignment to external noise [5, 15, 30, 44, 53], implying that external noise directly disturbs the coherence of a flock and that diverse velocities can only result from external perturbation. The relationship between the mechanism of establishing order and noise can be changed from a conflict relationship to a cooperative relationship by taking inherent noise into account.

Because inherent noise reveals the probability distribution of actions, an interaction is a choice from potential actions. If such potential actions were independent of each other, it would be very difficult for an individual to choose an action resulting in collective behavior. This implies that inherent noise can be regarded as external noise. Otherwise (i.e., if the potential actions of individuals were strongly correlated), an individual could easily choose an action to cooperate with collective behavior, suggesting that an individual could anticipate flock-mates' behavior to some extent. Anticipation has been addressed in hierarchical biological systems, such as the metabolic repair system [46]. In this context, anticipation is an optimistic prediction that is equipped with expectation. Avoiding a traffic jam also requires the idea of anticipation, which is nothing more than a swarm avoiding collisions [17, 31, 51].

Our research group explores new ideas on anticipation with the interaction between the present and the future under the asynchronous updating process [22, 23, 24, 35, 36, 37, 38, 41, 47]. Although the interaction between the present and the future seems not to be a real physical phenomenon, it implies that the individual in a swarm can know the tendencies of the flock-mates' behavior. Because most of animals show typical behaviors proper to their species, it is natural that animals have knowledge of repetitive behaviors shown by their flock-mates. Asynchronous updating is also natural because perception and/or sensory information is frequently delayed and there is a wide variety with respect to the delay time within species. The idea of anticipation in collective motion is also applied to morphogenesis [25]. Recently the idea of anticipation was replaced by Bayesian inference under the condition of which the likelihood of the hypothesis is also changed (i.e. inverse Bayesian), and the wide variety of updating timing is described by using probability [26].

Independent of our ideas on anticipation, some researchers also have begun to focus on the significance of anticipation in an animal group. Morin et al. [34] implements anticipation by the knowledge of the tendency of the flock-mates' behavior in terms of direction. Gerlee et al. [19] explores this stance in a similar manner to us and introduces the interaction between the present and future time. On the other hand, Piwowarczyk et al. [43] introduces wide variety of sensory delays in a swarm and shows us how the delay is intrinsic as noise plays a role in dynamic collective motion.

Previous studies have shown that a correlated domain in a bird flock; scales with the flock size [10]. It has also been shown that the correlation function decays linearly with the distance rescaled by the correlation length, which is the distance with the zero-correlation function [10]. The positional information in a body results from an intrinsic adjusting mechanism that may be driven by inherent perturbations. It has also been suggested that inherent turbulence could play an essential role in collective motion [7, 49, 55]. Thus, a flock or school can be a scale-free, parallel-distributed system driven by inherent perturbation.

In this context, we focused on the soldier crab, *Mictyris guinotae*, as a model organism to investigate a swarm driven by internal noise and/or anticipation. These crabs wander in lagoons at low tide and can generate a swarm. Usually, a swarm composed of crabs crawls near a tidal pool and suddenly enters it en masse. There are various internal flows moving within a swarm, although the boundary of the swarm is clear and conspicuous. Inherent noise and/or anticipation may thus cooperate to result in collective behavior and dynamic changes in global motion.

Here, we show that inherently generated noise can positively contribute to coherent collective behavior, revealing a scale-free correlation in both a computer model and real kinetic data from soldier crab swarming. First, we describe the behavior of solider crabs, dependent on a field survey and show a model in which an individual has an inherent perturbation and the capacity for mutual anticipation. Second, we illustrate typical water-crossing behavior observed in real soldier crabs. Third, we show how our model can approximate real kinetic data from soldier crabs in terms of the strength of the coherence and the inherent perturbation. We also show that both our model and real kinetic data reveals a scale-free correlation that is the same as that which has been described for starlings. Additionally, we implement an hourglass made of soldier crabs that exhibits periodic oscillation. Finally, we elaborate on the significance of inherent noise and/or mutual anticipation in highlighting the difference between robustness and stability.

2. Model

2.1 Overview

Based on numerous field surveys of *M. guinotae*, we have found the following observations to be characteristic of soldier crab swarming behavior [6, 42, 48]: (i) a swarm moving in a lagoon has internal turbulence (i.e., crabs exhibit different velocities within a swarm), suggesting that inherent noise is generated within a swarm; (ii) turbulent motion results in part of the swarm becoming highly concentrated, and this part enters and crosses the water because of the effect of the group; and (iii) individuals in other parts of the swarm then follow their predecessors.

Previous models of flocks or swarms have been based on local interactions between agents in their own neighborhood. In these models, the force that generated the collective behavior or order conflicts with noise, and the dynamic turbulent behavior of a flock or school could result from the balance of generating order and noise. The larger the influence of noise, less coherent the structure of the generated flock or school. In contrast, inherent noise or internal turbulence can contribute to collective behavior in the case of soldier crabs.

How can inherent noise contribute to collective behavior? If an individual chooses his own action from various potential actions and acts independently from the others, coherent collective behavior cannot be expected. In fact, a swarm could give rise to chaos. The coexistence of diverse actions is driven by inherent noise, and the coherent motions of a swarm needs specific interactions to generate coherence while allowing individual diversity. This has been addressed by anticipatory systems [46]. Because anticipation is considered to be a hopeful prediction, it requires a kind of sociality to achieve what is predicted. Such a hopeful prediction is possible in a society where an individual can perceive the disposition of its neighbors, and the individuals can move asynchronously. This assumption is reasonable because even a human can perceive the disposition of his or her neighbors in a crowded walkway to avoid collision [27]. Furthermore, soldier crabs have 360-degree eyesight due to high resolution in the equatorial region of their eyes [40, 56], making them sensitive to their neighbors position.

To depict the behaviors of soldier crabs (i-iii) in a model based on mutual anticipation, we introduced many potential transitions for an individual that could anticipate the movements of others within the swarm (Fig. 1). An individual has its own basic vector (velocity) accompanied by P, the number of potential transitions in a range restricted by the angle, α. The individual located at the center of the neighborhood chooses one of the potential transitions depending on mutual anticipation that is defined by following.

In Fig. 1, in a neighborhood, P number of potential transitions (black arrow) are derived from a basic vector (red arrow) having the maximal

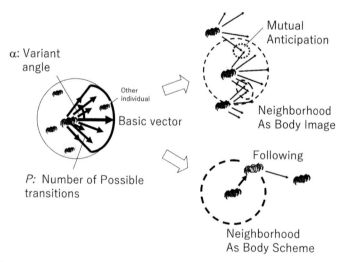

Fig. 1. Schematic diagram of our model featuring mutual anticipation.

angle of the angle, α, on one hand, and other flock-mates located in the neighborhood are referred by the individual, on the other hand. If potential transitions have overlapping targets, mutual anticipation (i.e., potential resonance) occurs, and an individual can move to the site of the overlap in the asynchronous updating fashion. Otherwise, an individual follows its predecessor.

We assume here that an individual can see the overlapping sites of targets of potential transitions of some individuals. This assumption implies that an individual feels the atmosphere in the neighborhood and/ or has the knowledge of the tendency of other individuals' (flock-mates') behaviors to some extent. Then, the individuals attempts to move to one of the overlapping sites in the neighborhood. If many individuals rush to the same overlapping sites, they collide at the site. To avoid such collisions, we introduce asynchronous updating of the move to the overlapping site. Imagine that one such overlapping site is shared by three individuals A, B and C. It implies that the individuals A, B and C can move the overlapping site. The updating order is randomly given, and for instance, B, C and A, and then the individual B first moves to the overlapping site, and the individual C secondly moves to the other overlapping site if possible. Finally, the individual A moves to the other overlapping site if possible.

In Fig. 2 individual crabs are represented by blue cells and potential transitions are represented by red arrows. The principal vector is especially represented by thick red arrow. First, velocity matching is applied to the

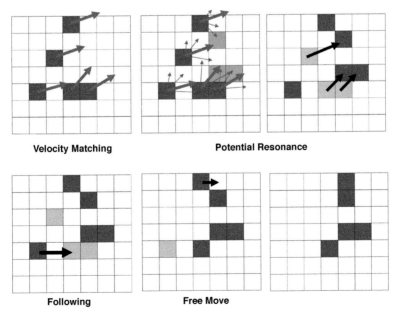

Velocity Matching **Potential Resonance**

Following **Free Move**

Fig. 2. Illustration of a transition of a swarm in lattice space.

basic vectors (upper left). Next, the mutual anticipation is estimated. The amount of overlap between targets of potential transitions is calculated, and sites with an overlap larger than 1 (threshold) are obtained (pink site). For example, the popularity of the most upper pink site is 2 (upper center). An individual whose potential transitions reach certain popular sites moves to the most popular site (black arrow in the upper right). After that, followers (lower left) and free wanderers (lower center) move, which results in the final distribution (lower right).

If there is no overlapping site, the individual chooses one of the potential transitions which have no overlapping site, we call this a free move, or the individual moves to the vacant site which is occupied by other individual at the preceding time step, we call this following. In this sense, the neighborhood has double meaning with respect to the analogy of body and neighborhood. While the neighborhood can be regarded as the extended body, the body has double meaning; the one as a sensory field which is called body image and the other as an object moved by the owner which is called body scheme. In our model, the neighborhood in which mutual anticipation is detected has the meaning of body image, and the neighborhood whose modification due to the vacant site is detected and can be moved has the meaning of body scheme (Fig. 1).

Figure 2 shows some examples of mutual anticipation, following and the free moves of our model represented in a lattice space. If the targets of some potential transitions overlap at a certain site, the number of transitions reaching that site is counted as the "popularity" (Fig. 2). If there are some sites where the popularity exceeds a threshold for the targets of an individual's transitions, then the individual moves to the site with the highest popularity. If several individuals intend to move to the same site, one individual is randomly chosen to move there, and the others move to the site with the second highest popularity. This rule represents an individual's mutual anticipation. If there is no site where the popularity exceeds the threshold value within the neighborhood of an individual and if others in the neighborhood move due to mutual anticipation, then the individual will move to occupy the absent site generated by mutual anticipation. Namely, the individual follows its predecessor. However, if an individual's movements are not based on mutual anticipation or the actions of a predecessor, then it moves in the direction of a randomly chosen potential transition and is called a free wanderer (Fig. 2).

2.2 Details of the Basic Model

The following is the detailed definition of our model [22, 23, 25, 35, 36, 38]. The model consists of N individuals moving within a discrete time and discrete $S \times S$ space, where $S = \{1, 2, ..., s_{MAX}\}$. The discrete space allows us to simplify the rule which was also applied to the previous flock model

[52]. The location of the k-th individual at the t-th step is given by $P(k, t)$ $= (x, y)$, where $x \in S$, $y \in S$, $k \in K = \{1, 2, \dots, N\}$, and the boundary condition is given in a wrapped fashion.

Each k-th individual at the t-th step has P number of potential vectors $v(k, t; i)$, with $i \in I = \{0, 1, \dots, P-1\}$. If $i=0$, the vector $v(k, t; 0)$, called the principal vector or basic vector, is represented by the angle $\theta_{k,t}$, such that $v(k, t; 0) = (Int(L\cos\theta_{k,t}), Int(L\sin\theta_{k,t}))$. For any real number x, $Int(x)$ represents the integer X such that $X \leq x < X+1$. L is the length of the principal vector. Because of the wrapped boundary condition, $X \in S$. If $i \neq 0$, the vector is defined using a random value, η_i, selected with equal probability from $[0.0, 1.0]$, and a random value (radian), ξ_i, selected with equal probability from $[-\alpha\pi, \alpha\pi]$. It is expressed as

$$v(k, t; i) = (Int(L\eta_i \cos(\theta_{k,t} + \xi_i)), Int(L\eta_i \sin(\theta_{k,t} + \xi_i))). \tag{1}$$

The principal vector $v(k, t; 0)$ is a special case where $\eta_0 = 1.0$ and $\xi_0 = 0.0$. For each $v(k, t; i)$, the target of the vector is represented by $\tau(k, t; i)$, such that

$$\tau(k, t; i) = P(k, t) + v(k, t; i). \tag{2}$$

To implement mutual anticipation, we define the popularity of the targets of the vectors. The popularity is defined for each site at the t-th step, (x, y) with $x \in S$, $y \in S$, by

$$\zeta(x, y; t) = |\{\tau(k, t; i), k \in K, i \in I \mid \tau(k, t; i) = (x, y), {}^{\forall}(k \in K)P(k, t) \neq (x, y)\}|. \tag{3}$$

The symbol $|S|$ with a set S represents the number of elements of S. The popularity of each site implies the number of transitions whose target reaches the site and is not occupied by a crab.

By using the popularity, the transition is asynchronously updated. To implement asynchronous updating, we define $\omega(x, y; t) \in \{0, 1\}$ to check whether the site has been updated or not. Before updating the location, for any (x, y) at the t-th step, we set $\omega(x, y; t) = 0$. After updating the site (i.e., an agent has moved from the site, which will be vacant in the next step), $\omega(x, y; t) = 1$.

At any time-step t, the updating agent is randomly chosen from $\{(x, y) \in S \times S \mid \omega(x, y; t) = 0\}$ where the agent here is represented by the co-ordinate at which the agent stays at t-th step. The individuals are thus asynchronously updated. The order of updating is randomly determined and independent of the number of individuals, k.

If there exists an $i \in I$ such that $\zeta(\tau(k, t; i)) \geq 2$, then the next site for the k-th individual is defined by

$$P(k, t) = \tau(k, t; s), \tag{4}$$

where s satisfies the condition such that if for some $i \in I$, $\zeta(\tau(k, t; s)) \geq \zeta(\tau(k, t; i))$, $i = s$. It implies that the s-th potential transition can reach the maximal popular site for the updating agent. Thus, there can be some maximal popular sites. If so, one of them is randomly chosen. Note that equation (4) is not expressed by $P(k, t+1) = \tau(k, t; s)$. This is because the updating agent searches for the most popular site in vacant sites asynchronously. For the next updating agent, the site $\tau(k, t; s)$ is not a vacant site. Since the site is removed in $^{\forall}(k \in K)P(k, t) \neq (x, y)$ in the equation (3), $P(k, t) = \tau(k, t; s)$ is removed for the next updating agent.

This transition is shown as mutual anticipation in Fig. 2. In other words, the individual moves to the target of its own potential vector that has the maximal popularity.

Because updating is asynchronous, the set of sites updated by Equation (4) gradually grows. An individual that does not satisfy the condition corresponding to Equation (4) moves to a vacant site in the follower neighborhood, N_f whose radii is larger than the length of the basic vector. This move is called, following and is defined by

$$P(k, t) = Rd\{(x, y) \in N_f | \omega(x, y; t)=1\}, \tag{5}$$

where $Rd\{-\}$ represents an element randomly chosen from a set $\{-\}$. Note that $\omega(x, y; t)=1$ implies that the site (x, y) represents the site previously occupied by the updated agent. A following crab is presented in the lower left diagram in Fig. 2. If the individual moves neither by Equation (4) nor by Equation (5) (i.e., there is neither a popular site nor a vacant site that is previously occupied by other agents), it moves by

$$P(k, t) = Rd\{\tau(k, t; i) | ^{\forall}(j \in K)P(j, t) \neq \tau(k, t; i)\}. \tag{6}$$

It implies the target of potential transitions, $\tau(k, t; i)$ is vacant. This transition is shown as a free move in Fig. 2.

After all the agents are updated and moved to the new sites, time step is updated by;

$$P(k, t+1) = P(k, t). \tag{7}$$

It implies that t is not the moment but duration in which all agents are updated asynchronously, and that beyond the discrete time t asynchronous updating cannot occur.

Finally, the principal vectors are locally matched with each other in the neighborhood through velocity matching, M. This matching operation is expressed as

$$\theta_{k,t+1} = <\theta_{k,t}>_M. \tag{8}$$

The bracket with M in equation (8) represents the average velocity direction in the neighborhood M. This process is first depicted in Fig. 2.

In our models in [22, 23, 35, 36] , velocity matching is involved with the model, however the model which appeared in Gunji et al. [24] and Murakami et al. [38] does not contain the velocity matching while any other parts of the model are the same. The model without velocity matching can show a robust and dense swarm moving in the ballistic movement.

The parameters in our model are as follows: L, the length of the principal vector; P, the number of potential vectors; α, the angle derived from the principal vector; $R(N_f)$, the radii of the follower neighborhood; and $R(M)$, the radii of the neighborhood of velocity matching. The neighborhood with radius R of a site (cell) is defined by a square consisting of $(2R+1)^2$ cells in which the cell is centered. Depending on parameters P and α, the behavior of a swarm changes. The larger P is, the more robust and dense the swarm is, as discussed later.

2.3 Water-Crossing Behavior

In a simulation introducing a tidal pool, we defined a specific area $U_p \subseteq S \times S$ in which the condition allowing mutual anticipation (Equation (4)) is replaced by

$$\zeta(\tau(k,\,t;\,i)) \geq c. \tag{9}$$

The value c is an integer larger than two. Because $c > 2$, it is more difficult for individuals to pass through the area U_p, which mimics a tidal pool that an individual soldier crab does not enter. Only by introducing the specific area U_p can we simulate the water-crossing behavior observed in Murakami et al. [36].

2.4 Hourglass Condition

As described below, we encapsulated a swarm of soldier crabs in a container and implemented an hourglass made of crabs. To simulate the hourglass behavior of crabs, we gave the model crabs a tendency to walk along the wall. The hourglass situation was set as follows. We first defined the wall state for any lattice (x, y) such that $w(x, y) = 1$ if the site is a wall state and zero otherwise. In the hourglass situation, an individual can be located only at the site where $w(x, y) = 0$. The angle of tangential direction is defined for each wall state site (x, y) and is represented by $\theta_w(x, y)$. The tendency to walk along the wall is defined by

$$\theta_{k,t} = Rd\{\beta,\ \beta+\pi\} \tag{10a}$$

$$\beta = Rd\{\theta_w(x,\,y) \mid d(P(k,\,t),\,(x,\,y)) \leq d(P(k,\,t),\,(u,\,v)),\ w(x,\,y) = w(u,\,v) = 1\}, \tag{10b}$$

for $(x, y),\ (u, v) \in N_W$, where $d((p, q), (x, y))$ represents the metric distance between two sites, (p, q) and (x, y), and N_W represents the neighborhood of wall-monitoring for each individual. If an individual is close to the wall based on N_W, the individual's velocity, $\theta_{k,t}$, becomes parallel to the tangential direction of the wall. After this operation, velocity matching is applied to all individuals. Only by introducing Equations (10a) and

(10b) can individuals close to the wall walk along the wall, while other individuals take a shortcut to the wall.

3. Materials and Methods

3.1 Location and Materials

We used the soldier crab, *M. guinotae,* as a model organism to investigate the collective behavior of animal groups. These crabs live in lagoons and form huge swarms composed of several hundred to hundreds of thousands of individuals [6, 48]. As described above, they wander in lagoons at low tide, with internal turbulent flows, avoiding tidal pools. Only if part of the swarm is highly concentrated will the swarm cross the pool (Fig. 3).

In Fig. 3 time proceeds from left to right, and from top to bottom. Red dots represent soldier crab individuals moving slowly. A pair of blue and yellow dots represents a fast moving individual, where blue and yellow represents a target and a source for an arrow (direction), respectively. The light gray area represents a tidal pool on the surface of a lagoon. Although a swarm first approaches the edge of the tidal pool, it wanders along the edge with inherent turbulence (top three photos). An individual or a small swarm often avoids entering water, which can explain this wandering. After this the middle part of a swarm becomes highly concentrated (center of the central row), this part of swarm enters and crosses the water.

Numerous field surveys of these crabs were conducted in Funaura Bay and Iriomote Island, Japan, in March and November 2008, August 2009, and June and October 2010. To obtain a visual record of crab behavior, namely two-dimensional movement on the lagoon, we used the video recording system shown in Fig. 4. Our system is composed of a DV camera and a camera tower with a width of 4,000 mm, a length of 2,000

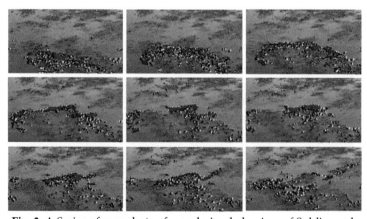

Fig. 3. A Series of snapshots of wandering behaviors of Soldier crabs

mm and a height of 825 mm. The system allows us to fix the DV camera in a downward position 805 mm above the lagoon surface. Fig. 5 shows the camera tower deployed in the field, and Fig. 6 shows a photo of soldier crabs taken by the camera.

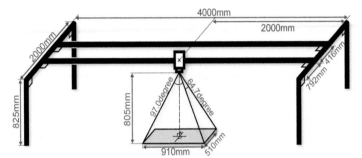

Fig. 4. Schematic of our camera tower. See text for the details.

Fig. 5. Our camera tower in Funaura Bay, on Iriomote Island, Okinawa, Japan.

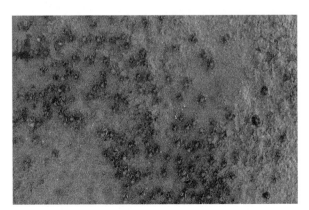

Fig. 6. Image of Soldier Crabs taken by the video camera.

The resultant kinetic data was used to compare the real swarm with our model simulation. While most of our visual data appeared in Murakami et al. [35, 36, 38] they are taken in a laboratory condition, and the data which has been presented in this paper is taken in natural field conditions.

3.2 Tracking

We attempted to determine each crab's position and track its movement by analyzing the recorded video. While it would be more efficient to automatically extract each crab's position and track it using image-processing technology, this is difficult to accomplish at present for the following reasons: (i) it is challenging to distinguish between the crabs and the background (i.e., the tideland) because the colors and patterns are similar and (ii) the visual condition of the tideland is not stable due to the effects of sunlight and the tides. Therefore, we developed software to record the position of a crab that the user has selected using a simple graphical user interface, making it possible to quickly record and track each crab's position as follows.

First, the user selects a video file to analyze, and the first frame image of the selected video is displayed automatically. By clicking the mouse on the shell of a crab in the image, the user can record the approximate location of the crab. Upon clicking, a blue square appears around the selected coordinate in the image. The center coordinate of the square is considered the crab's position. The user can then adjust the position of the blue square by dragging the mouse to achieve a more precise measurement of the crab's location. After this adjustment, the color of the square changes to red, and the crab's location is recorded (Fig. 7). It is possible to record

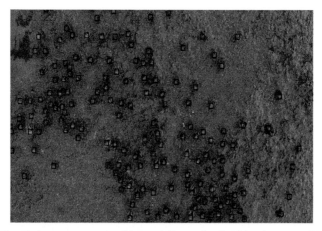

Fig. 7. Autonomous tracking soldier crabs. In this photo, multiple crabs are being tracked which is depicted by a red square.

the track of every crab by carrying out this operation for all images and all crabs. A red square remains around the crab after the user has finished analyzing it to prevent the same crab from being analyzed multiple times. This software was developed using Microsoft Visual Studio C# and OpenCVSharp (http://code.google.com/p/opencvsharp/), a cross-platform wrapper of OpenCV (Open Source Computer Vision Library, http://opencv.willowgarage.com/wiki/).

4. Results

We have implemented an asynchronous updating model in a lattice space coupled with velocity matching (VM) of the principal vectors (Fig. 2). By assuming at most one individual per lattice, the model implements collision avoidance (CA). The predecessor-following rule also implements flock centering (FC). These rules, VM, CA and FC, constitute BOIDS, a model that has recently been expanded to mimic more realistic characters (14, 28, 32, 33]. Thus, introducing mutual anticipation into our model is a natural extension of BOIDS and/or SPP.

Figure 8 shows us how various swarming patterns in the model are dependent on the parameters α and P. If $P=1$, our model mimics BOIDS because of the rule of VM. If $P=2$, multiple potential transitions break the collective motion. If P increases to larger than 2, mutual anticipation

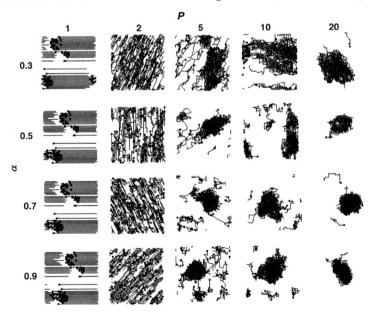

Fig. 8. How changing P and α affects the patterns of a swarm composed of 100 individuals in a 50×50 lattice: An individual is represented by a square and a trajectory tail, where $L=4$, $R(N_f)=2$ and $R(M)=3$.

contributes to swarm formation, especially if α is also large, which will cause polarization to decrease and trigger the formation of an oval swarm containing inherent turbulence. A coherent swarm will exhibit an amoeba-like motion, as turbulence within a swarm is very high ($P = 20$ and $\alpha > 0.5$ in Fig. 8). Because external perturbation cannot be distinguished from the inherent noise resulting from potential transitions, our model is robust against external perturbation due to mutual anticipation, which can be verified by simulation.

When we introduced a specific area in which the popularity threshold for triggering an individual's move was defined as higher than in other areas, the area (U_p, as defined by Equation (9)) could be interpreted as a tidal pool, which an individual soldier crab cannot enter [36]. Figure 9 shows the results of simulating a swarm crossing the water because of the effect of swarm density. After a period of transience, the individuals form a coherent and highly concentrated swarm. When a small swarm meets the peripheral area of a tidal pool, the swarm extends laterally and moves along the tidal pool area. Inherent turbulence can lead to the continuous modification of a swarm, bringing about a highly concentrated swarm. Soon after the highly concentrated swarm forms, it enters and crosses the water without an explicit leader or signals, similar with the movement of human crowds [18] and mimicking the behavior of real soldier crabs, as shown in Fig. 3.

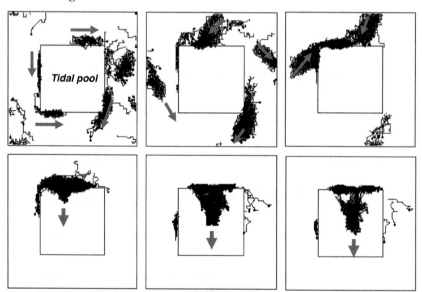

Fig. 9. Snapshot of a simulation demonstrating swarm behavior when crossing water. Red arrows represent the direction of the movement of the swarm. $L=4$, $R(N_f)=2$ and $R(M)=2$. Time goes from the top left to the top right, and then from the bottom left to the bottom right.

To estimate the effects of mutual anticipation, we plotted the density and polarization of a swarm against perturbation (Fig. 10). Polarization was defined by the length of the mean vector of the positional transitions of all agents. Density was defined by the normalized maximal number of agents in a swarm-size square. If the radii of a swarm can be approximated to L_s, the two dimensional space whose size is X by X is divided into X/L_s by X/L_s numbers of subspaces. Density is defined by the maximal number of agents in each subspace divided by the number of all agents.

This model without mutual anticipation, where an individual had only one transition (principal vector), was compared to the model incorporating mutual anticipation. In the model without mutual anticipation, the velocity of each agent was coupled to the external perturbation, chosen with uniform probability [-1.0, 1.0] (Fig. 10, left). Both the polarization and the density decreased as the perturbation increased, indicating that a high-density swarm is caused by high polarization.

Previous studies [52, 54] show that velocity matching can lead to the phase transition from the ordered state to the disordered (chaos) state, where the ordered state is expressed as high polarization. It is also known that the critical state at the edge of chaos has high performance of computation which can correspond to the various perturbed conditions. In other words, high performance is achieved only at the edge of chaos which requires detailed parameter tuning.

In contrast, in the mutual anticipation model, high density and low polarization were maintained under high internal perturbation, where the internal perturbation was defined as the number of potential transitions

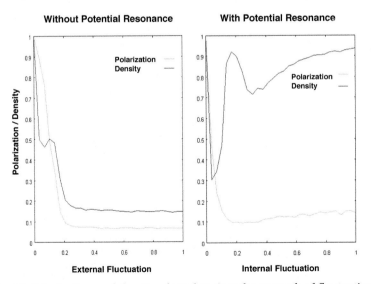

Fig. 10. Polarization and density plotted against the strength of fluctuation.

normalized to the maximal number of transitions (Fig. 10, right). The graph (Fig. 10 right) indicates that a highly dense swarm will be accompanied by strong inherent turbulence.

The diagram on the left shows the simulating results by the model composed of a single individual that has only a principal vector (i.e. the model without mutual anticipation). The transition of an individual is expressed only by velocity matching within a neighborhood coupled with external fluctuation. The diagram on the right shows the simulating results by the model with mutual anticipation. Polarization and density are plotted against internal fluctuation. Internal fluctuation is defined as the number of potential transitions normalized by the maximum number ($P=30$). $\alpha=0.7$, $L=4$, $R(N_f)=2$ and $R(M)=3$.

Our model shows that this tendency is independent of the angle α as shown in Fig. 11. It implies that highly dense swarm with low polarization is robust independent of the parameters.

A reconstruction of real wandering soldier crabs is shown in Fig. 12. It shows that the swarm consisting of about 150 crabs passed under our video camera system. Although the swarm moves in one direction overall, it is

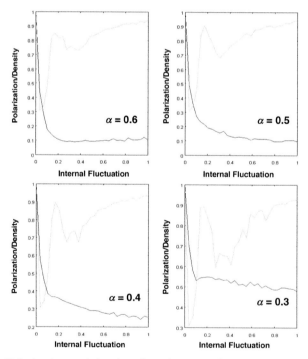

Fig. 11. Polarization and density plotted against the strength of fluctuation for the model with mutual anticipation: Each diagram shows polarization/density against internal fluctuation for the model with different angle α.

clear to see that individuals can wander within a swarm. We compared this real data with that generated by our model using a density/polarization plot (Fig. 13). A contour map shows the distribution of swarm density plotted against polarization. The plot in Fig. 14 illustrates how density/ polarization changes with the angle α. In each graph, P ranges from 10 to 20. When P is small, the density is low, independent of the polarization. As P increases, the density increases. When the angle is between 0.2 and 0.4 or 0.4 and 0.5 and P is large (>15), the density/polarization plots are distributed as shown in Fig. 13A. This distribution mimics that of a real soldier crab swarm, as presented in Fig. 13B.

Previous studies have shown that a correlated domain in a bird flock is scaled to flock size [10]. A correlated domain reveals a dynamic functional unit within a system. For example, if a snail extends a certain

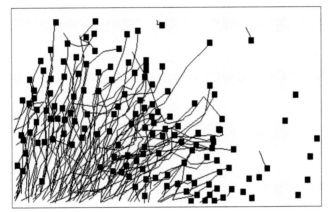

Fig. 12. Trajectory of the real soldier crabs: Each individual is represented by a black square with its trajectory.

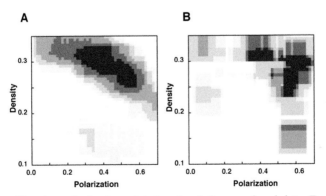

Fig. 13. Density-polarization plots for simulations and real data: Density plotted against polarization for model simulations over selected parameter values (A) and real data (B) with 10<P<20. L=4, $R(N_f)$=2 and $R(M)$=3.

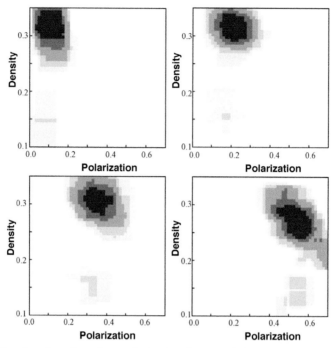

Fig. 14. Density plotted against polarization for a model simulation dependent on the angle α: $\alpha=0.5{\sim}0.6$ (upper left), $0.4{\sim}0.5$ (upper right), $0.3{\sim}0.4$ (lower left) or $0.2{\sim}0.3$ (lower right) with $10{<}P{<}20$. $L=4$, $R(N_f)=2$ and $R(M)=3$.

part of its body and moves other parts randomly, only the cells in the extended part are considered to be strongly correlated. Scale-free correlation implies that a functional unit exhibits a similar structure, regardless of the size [20]. Scale-free correlation is found only in starling flocks that can be approximated by SPP models. In such models [52, 53], the rules of alignment between individuals obey a rotational invariance, and a moving flock breaks the averaged polarization of the group. For this reason, it is expected that SPP models can explain the scale-free correlation observed in starling flocks. In contrast, a swarm of solider crabs seems to be linked by mutual anticipation, revealing internal and local turbulent flows within a moving group. Thus, the question that arises is whether scale-free correlation is also found in a swarm of soldier crabs and can be obtained from a model based on mutual anticipation.

A scale-free correlation was obtained for the distribution of the fluctuation vector $u_i = v_i - [1/N]\Sigma v_k$, where v_i is the velocity of the i-th individual and N is the number of individuals in a swarm. The correlation function $C(r)$ is defined by $C(r) = (\Sigma\, u_i \cdot u_i\, \delta(r - r_{ij}))/\Sigma\, \delta(r - r_{ij})$, where r is a distance, r_{ij} is the distance between the i-th and j-th individuals, and $\delta(x)$ is

Dirac's delta function. The correlation length ξ is defined by the distance with a zero-correlation function.

We applied this measure to a swarm of soldier crabs. For each snapshot of the transitions of the real swarm of soldier crabs (Fig. 15a), we obtained the correlation length (Fig. 15b). The correlation function was plotted against the rescaled distance normalized to the correlation length, as shown in Fig. 15c. We then compared the plots for the real crab swarm with the simulated swarm (Fig. 15d). The real kinetic soldier crab data exhibits linearity with the same slope as that of our model and that of a starling flock [10]. Thus, the internal structure reveals the scale-free, continuously changed "relative" positional information referring to a swarm as a unit. This scale-free correlation seems to be a consequence of asynchronous updating [21].

In our model, a solitary individual separated from its flock-mates embarks on a random walk because one potential transition is randomly chosen for each step. It makes sense that the potential transitions would represent inherent noise. Whenever individuals are highly concentrated, mutual anticipation can occur, and the inherent noise positively contributes to dense swarming. Therefore, even if individuals are exposed to high

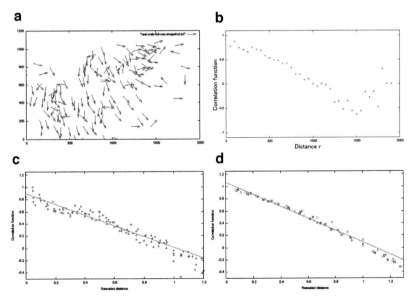

Fig. 15. Scale-free correlation, model simulation and real soldier crab kinetic data: (a) Snapshot of transitions for a real soldier crab. (b) Correlation function against distance, for real kinetic data of soldier crabs. It also shows how to obtain the correlated length with zero crossing of the correlation function. (c) Correlation function against rescaled distance, for real kinetic data of soldier crabs. (d) Correlation function with plotted against rescaled distance normalized by correlated length, for our model simulation. $L=4$, $R(N_f)=2$ and $R(M)=2$.

external perturbations, the perturbed transitions cannot be distinguished from the inherent noise, and a swarm resulting from mutual anticipation will be very robust to external perturbation. To enhance this robustness, we implemented a "crab hourglass."

Figure 16 shows a series of snapshots of a crab hourglass. In a container, the soldier crabs have a tendency of walking along the wall. First, most of

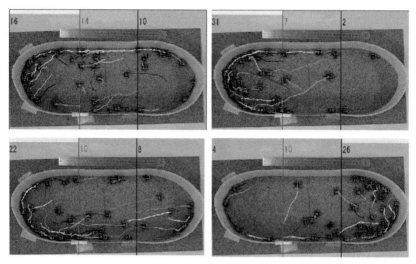

Fig. 16. Real soldier crab hourglass: Snapshots of real soldier crab hourglass. The area is divided into three areas, and the number of crabs in each area is counted and shown above. The substrate is made of cork board and supplies frictions.

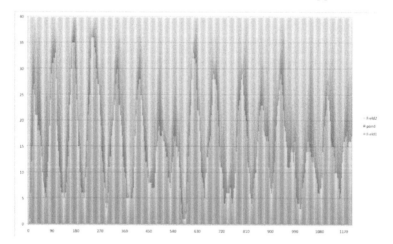

Fig. 17. Time series of the number of real soldier crabs in each area: Blue, red and green histogram shows the number of crabs in left, center and right area of the container, respectively.

the crabs moved to the left end, along the upper wall, resulting in a high concentration at that end. Next, the crabs moved to the right end. The number of individuals staying in the divided area oscillated periodically to some extent (Fig. 17). It is remarkable that this oscillation was maintained for about two hours.

Figure 18 shows snapshots from a simulation based on our model described above. It was assumed that an individual had a principal vector parallel to the tangent of the wall if it were close to the wall, where the direction chosen was of uniform probability to be clockwise or counterclockwise. The simulation results indicated that the swarm became highly concentrated at the left or right end and that the swarm rotated in a counterclockwise direction. Most of the individuals walked along the wall, although some took shortcuts. After a long period, the rotational direction changed from clockwise to counterclockwise and *vice versa.*

The number of individuals in a divided area oscillated regularly (Fig. 19), as well as in a real soldier crab hourglass. This oscillation mechanism is different from the periodic patterns of escape and pursuit behaviors observed in insect swarms [3, 45].

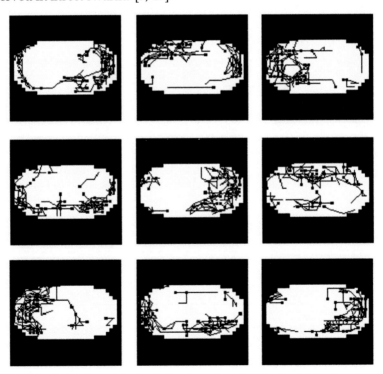

Fig. 18. Snapshots of model simulation based on the dual neighborhood system: An individual is represented by a black square with trajectory tail. Time proceeds from top to bottom and from left to right.

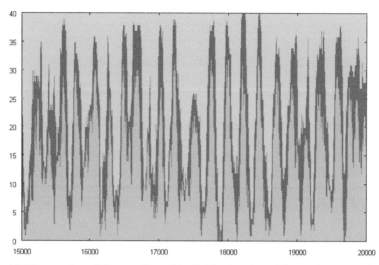

Fig. 19. Time series of the number of individuals of model simulation in divided area: Blue, red and green histograms shows the number of crabs in left center and right area of the container, respectively.

5. Discussion

In a real flock, school or herd, complex internal turbulence and vortexes within the group are observed [7, 49, 55]. Based on the field observations of soldier crabs, we confirmed that inherent noise positively contributes to the formation and maintenance of a conspicuously dense swarm. If inherent noise in the model is characterized by potential actions under synchronous updating, then the swarm will collapse since one action is randomly chosen at any instance. In contrast, if the diversity of actions can be used to anticipate other individuals' movement under asynchronous updating, this diversity results in a high probability of successful collective motion. Particularly, our simulation indicates that greater inherent noise [i.e., the larger P] results in the generation of a denser swarm (Fig. 8 and Fig. 10). Anticipation, therefore, is a key mechanism that produces inherent noise, and this noise, which could be expected to negatively affect the swarm, makes a positive contribution to collective motion. This anticipation may also be detected in some instances of biological collective motion [3, 45].

Mutual anticipation may be possible due to one crab's sensitivity to the movement of other individuals and the asynchronous motion of individuals in a swarm. Because of this sensitivity, crabs can detect the sites to which many individuals tend to move, and because of asynchronous updating, they can move in various directions without collision and overcrowding. This behavior can bring about internal turbulence in a dense and coherent swarm within a definite, conspicuous boundary. Due to the introduction of mutual anticipation, a swarm becomes robust to external perturbation. Our analysis using density/polarization plots suggests that a real soldier crab swarm can use inherent noise and anticipation (Figs. 13 and 14) to drive the phenomenon of robust collective movement.

The water crossing shown in Fig. 3 is a typical collective behavior resulting from inherent noise. Under normal conditions, crabs avoid entering water. Even if a crab is pushed by other mates in a swarm, the crab avoids the water or returns to land. However, when the inherent noise becomes very high and a high swarm concentration occurs along the marginal area of the water, the crabs rush into the water. Our model, based on mutual anticipation, can recapitulate this phenomenon only by introducing a high threshold value to trigger mutual anticipation. The water-crossing behavior can be explained by the explosion of the usually tamed inherent noise in the context of anticipation (Fig. 9).

Both our computer model and the real soldier crab data displayed scale-free correlations or positional information in a swarm (Fig. 15). It can result from mutual anticipation and/or inherent noise. This implies that inherent noise can reveal diversity within a swarm and can form a dynamic pattern consisting of correlated and uncorrelated domains. The dynamic, self-organizing nature of a swarm reveals that a swarm acts as a scale-free single body, embodying dynamic and parallel-distributed processing. These findings expand our understanding of the self-organizing collective behavior of a swarm by introducing the diversity of individuals.

The periodic oscillation of the crab hourglass also demonstrates a robust collective phenomenon driven by taming inherent noise (Fig. 16 and 17). The periodic oscillation of a concentrated swarm results from both positive and negative feedback, which is well known in biochemistry. Positive feedback alone, suggesting directed motion, cannot induce formation of an isolated and robust swarm. In contrast, the inhibition of polarization can make individuals wait for a delayed flockmate and can generate an isolated moving swarm. In our model, diverse potential transitions allow individuals in crowded parts of a swarm to move perpendicular to the averaged direction of motion (Figs. 18 and 19). As a result, individuals in the frontal area, which is usually crowded, can wait for the delayed individuals that follow their predecessors. Thus, taming inherent noise to obey mutual anticipation plays an essential role in creating oscillations.

We must distinguish robustness from stability in collective behavior. Stability is defined as the tendency to resist external perturbation. In this sense, noise is always in conflict with the stable mechanism of generating collective behavior. In contrast, mechanisms generating both order and noise can cooperate to result in robust collective behavior. Not only inherent but also external noise can contribute to maintaining robust collective behavior because external noise cannot be distinguished from inherent noise. The crab hourglass, illustrating crab water-crossing behavior and robust oscillation, confirmed that a collective behavior can be generated and maintained even in under perturbed conditions.

Here, "mutual anticipation" was not used in an anthropomorphic sense. Synchronous updating and/or the notion of time slices are assumed in most models to approximate biological phenomena. Clearly, biological processes asynchronously proceed to avoid turbulent behavior. If one assumes sensitivity to various physicochemical properties to generate the order of updating and asynchronous updating, the assumption of mutual anticipation is implied.

Recently our group also proposed the mechanism leading to collective behavior based on Bayesian inference under which the likelihood of hypotheses is perpetually changed [26]. It implements both the diversity of sensory delay and learning coupled with modification of learning targets. The former is relevant for asynchronous updating and the latter is relevant for anticipation. It can be regarded as another expression for mutual anticipation.

While inherent noise has been considered as an important property in collective biological phenomena, how inherent noise can contribute to robust order in a general context has not been investigated. In general, mutual anticipation is a promising candidate linking robust macroscopic collective behavior with microscopic inherent noise.

This work was supported by the Japan Society for the Promotion of Science, Scientific Research 17H01249.

References

1. M. Ballerini, V. Cabibbo, R. Candelier, E. Cisbani, I. Giardina, V. Lecomte, A. Orlamdi, G.P.A. Parisi, M. Viale and V. Zdravkovic. Empirical investigation of starling flocks: a benchmark study in collective animal behavior. Animal Behavior 76, 201–215 (2008a).
2. M. Ballerini, N. Cabibbo, R. Candelier, A. Cavagna, E. Cisbani, I. Giardina, V. Lecomte, A. Orlandi, G. Parisi, A. Procaccini, M. Viale and V. Zdravkovic. Interaction ruling animal collective behavior depends on topological rather than metric distance: evidence from a field study. PNAS 105, 1232–1237 (2008b).
3. S. Bazazi, J. Buhl, J.J. Hale, M.L. Anstey, G.A. Sword, S.J. Simpson and I.D. Couzin. Collective motion and cannibalism in locust marching bands. Current Biology 18, 735–739 (2008).
4. Ch. Becco, N. Vandewalle, J. Delcourt and P. Poncin. Experimental evidences of a structural and dynamical transition in fish school. Physica A 367, 487–493 (2006).

5. E. Bertin, M. Droz and G. Gr´egoire. Hydrodynamic equations for self-propelled particles: microscopic derivation and stability analysis. J Phys A: Math and Theor 42, 445001 (2009).

6. C. Bradshaw and T.P. Scoffin. Factors limiting distribution and activity patterns of the soldier crab Dotilla myctiroides in Phuket, South Thailand. Mar Biol 135, 83–87 (1999).

7. J. Buhl, D.J.T. Sumpter, I.D. Couzin, J.J. Hale, E. Despland, E.R. Miller and S.J. Simpson. From disorder to order in marching locusts. Science 312, 1402–1406 (2006).

8. V. Calenbuhr, L. Chretien, J.L Deneubourg and C. Detrain. A model for osmotropotactic orientation (II). J Theor Biol 158, 395–407 (1922).

9. C. Carere, S. Montanino, F. Moreschini, F. Zoratto, F. Chiarotti, D. Santucci and E. Alleva. Aerial flocking patterns of wintering starlings, *Sturnus vulgaris,* under different predation risk. Animal Behaviour 77, 101–107 (2009).

10. A. Cavagna, C. Cimarelli, I Giardina, G.S.R. Parisi, F. Stefanini and M. Viale. Scale-free correlation in the bird flocks. PNAS 107, 11865–11870 (2010).

11. I.D. Couzin. Collective minds. Nature 445, 715 (2007).

12. I.D. Couzin. Collective cognition in animal groups. Trends in Cog. Sc. 13, 36–43 (2008).

13. I.D. Couzin, J. Krause, N.R. Franks and S.A. Levin. Effective leadership and decision-making in animal groups on the move. Nature 433, 513–516 (2005).

14. I.D. Couzin, J. Krause, R. James, G.D. Ruxton and M.R. Franks. Collective memory and spatial sorting in animal groups. J Theor Biol 218, 1–11 (2002).

15. A. Czir´ok, E. Ben-Jacob, I. Cohen and T. Vicsek. Formation of complex bacterial colonies via self-generated vortices. Phyl Rev E 54(2), 1791–1801 (1966).

16. C. Detrain and J.L. Ddeneubourg. Self-organized structures in organism: do ants "behave" like molecules? Physics of Life Reviews 3, 162–187 (2006).

17. A. Doniec, S. Espi'e, R. Mandiau and S. Piechowiak. Dealing with multi-agent coordination by anticipation: application to the traffic simulation at junctions. Proc. EUMAS '05, 478–485 (2005).

18. J.R.G. Dyer, C.C. Ioannou, L.J. Morrell, D.P. Croft, I.D. Couzin, D.A. Waters and J. Krause. Consensus decision making in human crowds. Animal Behaviour 75, 461–470 (2008).

19. P. Gerlee, K. TunstrØm, T. Lundh and B. Wennberd. Impact of anticipation in dynamical systems. Phys. Rev. E 96, 062413 (2017).

20. Y.P. Gunji, T. Niizato, H. Murakami and I. Tani. Typ-Ken (an amalgam of type and token) drives Infosphere. Knowledge, Technology and Policy 23, 227–251 (2010).

21. P.Y. Gunji, T. Shirakawa, T. Niisato, M. Yamachiyo and I. Tani. An adaptive and robust biological network based on the vacant-particle transportation model. J. Theor. Biol. 272, 187–200 (2011a).

22. Y.P. Gunji, H. Murakami, T. Niizato, A. Adamatzky, Y. Nishiyama, K. Koichiro Enomoto, M. Toda, T. Moriyama, T. Matsui and K. Iizuka. Embodied swarming based on back propagation through time shows water-crossing, hour glass and logic-gate behavior. Advances in Artificial Life (Lenaerts, T. et al. eds.) pp. 294–301 (2011b).

23. Y.P. Gunji, Y. Nishiyama and A. Adamatzky. Robust soldier crab ball gate. Complex Systems 20, 94–104 (2011c).

24. Y.P. Gunji, H. Murakami, T. Niizato, K. Sonoda and A. Adamatzky. Passively active – actively passive: mutual anticipation in a communicative swarm. *In:* Plamen L. Simeonov, Leslie S. Smith and Andree C. Ehresmann (eds), Integral Biomathics: Tracing the Road to Reality, pp. 169–180. Springer, Verlag (2012).

25. Y.P. Gunji and R. Ono. Sociality of an agent during morphogenetic canalization: asynchronous updating with potential resonance. BioSystems 109, 420–429 (2012).

26. Y.P. Gunji, H. Murakami, T. Tomaru and V. Vasios. Inverse Bayesian inference in swarming behavior of soldier crabs. Philosophical Transaction of the Royal Society A 376, 20170370 (2018).

27. D. Helbing, F. Schweitzer, J. Keltsch and P. Mol☐ar. Active walker model for the formation of human and animal trail systems. Phys Rev E 56(3), 2527–2539 (1997).

28. C.K. Hemelrijk, H. Hindenbrandt, J. Reinders and E. Stamhuis. Emergence of oblong school shape: models and empirical data of fish. Ethology, 116, 1099–1112 (2010).

29. H. Hildenbrandt, C. Carere and C.K. Hemelrijk. Self-organized aerial displays of thousands of starlings: a model. Behavioral Ecology, 21(6), 1349–1359 (2010).

30. C. Huepe and M. Aldana. Intermittency and clustering in a system of self-driven particles. Phys Rev Let 92, 168701 (2004).

31. A. Klar and R. Wegener. Kinetic derivation of macroscopic anticipation model for vehicular traffic. SIAM J. Appl. Math. 60(5), 1749–1766 (2000).

32. H. Kunz and C.K. Hemelrijk. Artificial fish schools: collective effects of school size, body size, and body form. Artificial Life 9, 237–253 (2003).

33. R. Lukeman, Y.-X. Li and L. Edelstein-Keshet. Inferring individual rules from collective behavior. PNAS 107, 12576–12580 (2010).

34. A. Morin, J.B. Caussin, C. Eloy and. D. Bartolo. Collective motion with anticipation: flocking, spinning, and swarming. Phys. Rev. E91, 012134 (2015).

35. H. Murakami, T. Niizato and Y.P. Gunji. A model of a scale-free proportion based on mutual anticipation. International Journal of Artificial Life Research, 3(1), 34–44 (2012).

36. H. Murakami, T. Tomaru, Y. Nishiyama, T. Moriyama, T. Niizato and Y.P. Gunji. Emergent runaway into an avoidance area in a swarm of soldier crabs. PLoS ONE 9(5): e97870, doi:10.1371/journal.pone.0097870 (2014).

37. H. Murakami, T. Niizato, T. Tomaru, Y. Nishiyama, Y.P. Gunji. Inherent noise appears as a Lavy walk in fish schools. Scientific Report 5, 10605 (2015).

38. H. Murakami, T. Niizato and Y.P. Gunji. Emergence of a coherent and cohesive swarm based on mutual anticipation. Scientific Reports 7, 46447 (2017).

39. M. Nagy, Z. Akos, D. Biro and T. Vicsek. Hierarchical group dynamics in pigeon flocks. Nature 464, 890–893 (2010).

40. H.O. Nalbach, G. Nalbach and L. Forzin. Visual control of eye-stalk orientation in crabs: vertical optokinetics, visual fixation of the horizon, and eye design. Journal of Comparative Physiology A 165, 577–587 (1989).

41. T. Niizato and Y.P. Gunji. Fluctuation-driven flocking movement in three dimensions and scale-free correlation. PlosOne 7, e35615 (2012).

42. J.F. Peter, D. Hsi-Te Shih and B.K.K. Chan. A new species of *Mictyris guinotae* (Decapods, Brachyura, Mictiridae) from Ryukyu Island, Japan. Crustaceana Monographs 11, 83–105 (2010).

43. R. Piwowarczyk, M. Selin, T. Ihle and G. Volpe. Influence of sensorial delay on clustering and swarming. Phys. Rev. E 100, 012607 (2019).

44. C.W. Reynolds. Flocks, herds, and schools: a distributed behavioral model. Computer Graphics 21(4), 25–34 (1987).

45. P. Romanczuk, I.D. Couzin and L. Schimansky-Geier. Collective motion due to individual escape and pursuit response. Phy Rev Let 102, 010602 (2009).

46. R. Rosen. Anticipatory Systems – Philosophical, Mathematical and Methodological Foundations. Pergamon Press (1985).

47. T. Sakiyama and Y.P. Gunji. The Müller-Lyer Illusion in ant foraging. PLoS ONE 8(12): e81714 (2013).

48. J.T. Shih. Population-densities and annual activities of *Mictyris brevidactylus* (Stimpson, 1858) in the Tanshui mangrove swamp of northern Taiwan. Zool. Stud. 34, 96–105 (1995).

49. S.J. Simpson, G.A. Sword, P.D. Lorch and I.D. Couzin. Cannibalism crickets on a forced march for protein and salt. PNAS 103, 4152–4156 (2006).

50. D.J.T. Sumpter. Collective Animal Behavior. Princeton, Univ. Press, Princeton (2010).

51. M. Treiber, A. Kesting and D. Helbing. Delays, inaccuracies and anticipation in microscopic traffic models. Physics A 360, 71–88 (2006).

52. T. Vicsek. Fluctuations and Scaling in Biology. Oxford, Oxford Univ. Press (2001).

53. Y. Vicsek, A. Czirok, E. Ben-Jacob, I. Cohen and O. Shochet. Novel type of phase transition in a system of self-driven particles. Phys Rev Let 75, 1226–1229 (1995).

54. T. Vicsek and A. Zafiris. Collective motion. Arxiv preprint arXiv:1010.5017 (2010). arxiv.org.

55. C.A. Yates, R. Erban, C. Escudero, I.D. Couzin, J. Buhl, I.G. Kevrekidis, P.K. Maini and D.J.T. Sumpterh. Inherent noise can facilitate coherence in collective swarm motion. PNAS 106, 5464–5469 (2009).

56. J. Zeil, G. Nalbach and H.O. Nalbach. Eyes, eye stalks and the visual world of semi-terrestrial crabs. Journal of Comparative Physiology A 159, 801–811 (1986).

5

Swarm Intelligence in Cybersecurity

Cong Truong Thanh*, Quoc Bao Diep and Ivan Zelinka

VSB-Technical University of Ostrava Department of Computer Science
Faculty of Electrical Engineering and Computer Science Ostrava, Czech Republic

1. Introduction

Today, cybersecurity has become an essential part for computer systems and infrastructures with a concentrate on the defense of valuable resource stored on those systems from malicious actors who want to steal, damage, destroy or forbid access to it. Contemporary with the development of technology, cyber-threats are becoming more sophisticated and automated. Cybercriminals are always changing their methods, making attacks challenging to predict and prevent.

Over the years, scientists have investigated a considerable number of methods to protect information systems from unauthorized access and unauthorized use. Such techniques may involve proper implementation of security mechanisms such as passwords, encryption, access control lists as well as complicated security protocols.

Nevertheless, conventional security systems that use rules and signatures, have become ineffective against flexible, continually evolving cyber-attacks. Furthermore, current cyber defense systems involve humans at multiple levels, which may cause the information flow slow and asynchronous. As a consequence, such system is unavailable to adapt to the speeds of cyber threats.

Thus, we need advanced approaches such as applying swarm-based intelligence techniques that provide flexibility and will have a learning capability to assist humans in combating cybercrimes.

The main aim of this study is to present the capability of adopting SI techniques for combating cybercrimes, to demonstrate how these techniques can be an effective tool for combating cyber-attacks. Furthermore, this study suggests directions for future research.

---

*Corresponding author: cong.thanh.truong.st@vsb.cz

The rest of this paper proceeds as follows. Section 2 introduces the concept of swarm intelligence. It is followed by Section 3, which provides insight, categorizes and surveys several SI-based approaches for used in combating against cyber-threats. Then, section 4 discusses the limitations of existing approaches as well as points out the scope for future work in the field. Finally, section 5 concludes the chapter.

2. Overview of Swarm Intelligence Techniques

Nature's inspiration has always helped humanity to solve real life problems. Over the years, the application of bio-inspired techniques have been seen in various areas ranging from computer science, engineering, economics, medicine, as well as the social sciences. Among those biologically-inspired techniques, SI is a research direction that has attracted scientists in the recent past. This domain focuses on designing robust systems (or principles) that can operate efficiently and intelligently under threat and catastrophic conditions without centralized control [10]. SI-based algorithms have emerged as nature-inspired algorithms that are capable of providing quick, robust, and inexpensive solutions to various complex problems [14, 40].

The expression "Swarm Intelligence" was first introduced by Beni and Wang in the context of the cellular robotics system [8]. This term refers to the collective behavior of agents in nature such as ant colonies, honey bees, fireflies, and bird flocks. These agents in the swarm collaborate with one another to achieve tasks necessary for their survival. Generally speaking, SI is a population-based system that comprises of many individuals in that they communicate locally with each other and with their environment. The individuals interact with each other via direct or indirect techniques to solve problems [38]. Direct interaction may occur through direct contact, such as ants touching each other with their antenna or through audio and visual techniques. On the other hand, indirect interaction, which is referred to as stigmergy, means that the individuals communicate through the environment, such as the product pheromone ants use for exchanging information [18]. SI can therefore be defined as a branch of Artificial Intelligence (AI) which is used to model the collective activities of social agents in nature.

To date, many SI algorithms have been proposed in literature and successfully applied in practice, including function optimization problems, finding optimal routes, scheduling, structural optimization, and image analysis [22]. Examples of SI algorithms are: Ant Colony Optimization [18, 19, 20], Particle Swarm Optimization [21, 30, 31], Artificial Fish Swarm [34], Artificial Bee Colony [28, 29], Self-Organizing Migrating Algorithm [17, 53], Bacterial Foraging [41], Cat Swarm Optimization [15], Glowworm Swarm Optimization [32], Firefly Algorithm [51], Bat Algorithm [52],

Grey Wolf Optimizer [37], and Whale Optimization Algorithm [36]. The most well known and very basic one is Ant Colony Optimization (ACO), followed by Particle Swarm Optimization (PSO) and Self-Organizing Migrating Algorithm (SOMA). As a consequence, these methods were utilized in a variety of problem domains as well as being applied in cybersecurity.

2.1 Ant Colony Optimization

In the natural world, ants are highly social creatures. They interact with each other and collaborate to solve complex problems such as finding a path to their food source from their nest. The ant colony conducts their food search operation by first sending exploratory ants in random directions. When the first ants discover food, they travel back to their nest while releasing a chemical called pheromone along the path. The other ants might get attracted by the pheromone and follow this path. Paths with a higher pheromone density have a higher chance of being chosen. Nevertheless, over time, the pheromone on the path will evaporate, thus reducing its concentration. Shorter roads will have a higher pheromone concentration compared to longer paths because more ants will pass. Further, if the pheromone is wholly evaporated, ants will not follow that path anymore. As a consequence, when ants find the best route to their food source, other ants will also move along and have positive feedback on that path, which leads to the following ants to move on a single route.

This type of behavior of ants to reach their food source through the shortest distance lead to the development of sophisticated optimization methods and algorithms. Dorigo [18, 19, 20] adopted this natural process to formulate ant colony optimization (ACO) algorithm. ACO has been applied for solving many combinatorial optimization problems such as scheduling, probabilistic Traveling Salesman Problem, assembly line balancing, graph coloring, and data mining problems.

The transition probability (p_{ij}) is the most crucial factor of the ACO algorithm. This probability represents the chance of an ant to travel along the path from i to j, which can be defined by the Eq. 1:

$$p_{ij}^k = \frac{(\tau_{ij}^\alpha)(\eta_{ij}^\beta)}{\sum_{l \in N_i^k}(\tau_{il}^\alpha)(\eta_{il}^\beta)}, j \in N_i^k \tag{1}$$

where:

p_{ij}^k : is the probability of the kth ant to move from node i to node j,

N_i^k : is the set of nodes in the neighborhood of the kth ant in the ith node,

τ_{ij}^α : is the the amount or intensity of the pheromone on the edge ij or desirability of a path,

η_{ij}^β : is the visibility of the path between i and j

α, β : weight parameters that control the influence of τ_{ij} and η_{ij} correspondingly.

When the ant has finished its job, it has to return to the nest. On the way back, it deploys a certain amount of pheromone on the traveled path. The amount of pheromone deposited on a path is illustrated in Eq. 2

$$\delta\tau_{ij}^k = \frac{Q}{L^k} \text{ if } (i, j) \in T^k \tag{2}$$

where:
Q : is a constant,
L^k : is the cost for the tour of kth ant.

When all the ants have completed their quest, the trails are updated by the following equation Eq. 3:

$$\tau_{ij} = (1 - \rho)\tau_{ij} + \sum_k \delta\tau_{ij}^k \tag{3}$$

where:
τ_{ij} : is the amount or intensity of the pheromone on the edge ij,
ρ : is the rate of evaporation, which controls the decay of pheromone.
$\delta\tau_{ij}^k$: is the amount of pheromone deposited by k^{th} ant.

2.2 Particle Swarm Optimization

Early studies [27, 42] indicated that in nature, a flock (swarm) of birds or animals is established when individuals follow simple rules:

- Collision avoidance; individuals avoid neighbor mates by adjusting their position.
- Velocity matching; individuals always synchronize their speed with their neighbor mates.
- Flock centering; members of the flock stay close to their mates.

The communication among individuals of a flock (or swarm), to share information, can be very beneficial to achieve their common goals such as raising the chance of discovering food.

Inspired by flocking behavior, Kennedy and Eberhart developed a heuristic optimization technique called Particle Swarm Optimization (PSO) [21, 30]. According to the PSO algorithm, a fitness function $f : \mathbb{R}^n \to \mathbb{R}$ is utilized to evaluate the quality of the current solution. The potential particles (solutions) are randomly distributed inside the hyperspace at positions $x_i \in \mathbb{R}^n$, each solution has the initial random velocity $v^i \in \mathbb{R}^n$. The potential particles traverse throughout the hyperspace, and at each step, their positions are evaluated according to the fitness function. The basic update rules for velocity and position illustrated in Eq. 4 and Eq. 5, respectively.

$$v_i(t+1) = \omega v_i(t) + c_1 r_1 (p_i - X_i) + c_2 r_2 (g - X_i) \tag{4}$$

$$x_i(t+1) = x_i + v_i(t+1) \tag{5}$$

where:

ω : the constant for inertia weight,

c_1, c_2 : the constants for acceleration,

r_1, r_2 : randomly generated numbers,

p_i : best personal position of the particle i,

g : global best position among all particles in the swarm,

X_i : the current position of particle i.

2.3 The Self-organizing Migrating Algorithm

The Self-Organizing Migrating Algorithm (SOMA) is based on the collaborative searching (migrating) of the area of all possible solutions (search area). During the search process, the individuals influenced by each other, which leads to creating or eliminating groups [17, 53].

In numerical optimization area, SOMA is applied to find globally optimal solutions. The process proceeded as follows. First, a population containing a given number of individuals is generated, in which each is a candidate solution to the problem. Next, the population is evaluated, and the best solution becomes the leader. The evaluation to find the leader is based on the competition – cooperation between these individuals, which is an essential feature of the swarm intelligent algorithm. This process repeats until satisfying the terminal condition. The detail of the whole process is described as the following.

The process begins with generating individuals randomly in the whole search space to form a population, as described in Eq. 6. In the initial population, each is a prospect solution to the problem. Then, the given fitness function will evaluate the population [17, 53].

$$P = x_j^{(lo)} + rand[0, 1](x_j^{(hi)} - x_j^{(lo)}) \tag{6}$$

where:

P : the initial population of the algorithm,

$x_j^{(lo)}$: the lowest boundary value,

$x_j^{(hi)}$: the highest boundary value,

$rand[0, 1]$: random number from 0 to 1.

In each migration circle, an individual with the best fitness value becomes the Leader. On the contrary, the others become the wandering individuals that will move step by step toward the Leader until reaching the *PathLength*. The *Step* determines the granularity of the migration process and the *PathLength* controls how far the individuals will jump.

Step and *PathLength* are prescribed, and fixed numbers in the algorithm. Eq. 7 describes the jumping process.

$$x_{n,j}^{ML+1} = x_{c,j}^{ML} + (x_{l,j}^{ML} - x_{c,j}^{ML})\, tPRT\ Vector_j \tag{7}$$

where:

$x_{n,j}^{ML+1}$: the new position in the next migration loop,

$x_{c,j}^{ML}$: the position in current migration loop,

$x_{l,j}^{ML}$: the leader position in current migration loop,
t : jumping step, from 0, by *Step*, to *P athLength*.

Before the jumping of an individual, a number for each dimension is randomly generated. Then this number is compared to the given *PRT* number. If it is larger than *PRT* , the *PRT Vector$_j$* of that jump is set to 0 and vice versa, *PRT Vector$_j$* is set to 1, as Eq. 8. This means that the individuals are moving in the $N - k$ dimensional subspace instead of moving directly to the Leader.

$$\text{if } rand_j < PRT;\ PRT\ Vector_j = 1;\ \text{else, } 0. \tag{8}$$

In the next migration circle, the Leader will be chosen again with another best individual, and the jumping processes continues to execute until the algorithm reaches the stopping criterion.

3. Swarm Intelligence Approaches for Combating Cyber-Threats

SI-based methods aim at solving complex problems with the use of multiple simple agents without the presence of a centralized authority. The agents collaborate with each other or with their environment to carry out difficult tasks. Because of their robustness, highly adaptive, and cost-efficient of SI algorithms, have been utilized in some areas of cybersecurity.

3.1 Malware Detection and Prevention

Malware is a general term for many types of malicious softwares such as virus, worm, trojan horse, exploits, botnet, retrovirus [48], and today it is a popular method of cyberattack. The malware's impact on digital society is enormous, therefore a considerable amount of research has been done to prevent and mitigate malware. Due to many common characteristics between malware and biological viruses, bio-inspired techniques have become a natural path to finding preventive solutions to malware problems.

In recent years, machine learning approaches have been used to improve the efficiency of malware detection. In literature, many studies have proposed applying the ML method to detect malware with high detection accuracy [50]. Additionally, SI-based techniques, which used for features optimization and optimizing the parameters, are proved to be capable of enhancing ML approaches.

The authors [2] proposed DyHAP, a hybrid method that is based on the adaptive neural fuzzy interface system and PSO to forecast the optimum parameters of mobile malware analysis. Moreover, the benign apps were downloaded from the Google Play store, and 1260 malware data samples were in the Malgenome datasets, and they captured the network patterns of 1,000 samples. According to their experiments, they accomplished optimized results by dealing with complicated parameters with RMSE: 0.43106, and R^2 : 0.7721. In the same manner, researchers in [6] propose a hybrid method that used the Information Gain (IG) and Pearson CorrCoef (PC) to rank the permissions and API permissions and API calls and PSO to optimize the parameters for Android malware detection. The datasets of malware and benign samples were gathered from various sources. The outcome of their approach showed that they archived an accuracy of 89% of malware detection.

In a later study, Ali et al. [5] provided a mining-based method for analysis and detection of obfuscated malicious software on the Windows environment. More precisely, the PSO algorithm was utilized for features selection and the Random forest (RF) for classification. The empirical results showed that the proposed method is efficient with an accuracy of 99.6% and outperformed the comparative algorithms in most of the experiments.

The authors in [45] used PSO to optimize the permission features in Android malware detection. In this study, they proposed to integrate PSO and Genetic Algorithm (GA) with the Random forest algorithm. According to the experiments, the utilization of PSO helped to increase the accuracy of classifying malware. Likewise, Bhattacharya et al. in [9] proposed to improve the PSO algorithm for feature selection method through a rough set of permissions to improve the performance of Android malware classification. The authors reported that their approach had a better performance over conventional feature selectors with the average accuracy from 87.68% to 97.86% depended on the dataset.

Another approach is in Ref. [46] that utilized the Whale Optimization Algorithm (WOA) technique for filtering Android malware. In this study, the WOA was used to extract the best features for three classification algorithms, namely Nave Bayes, J48 decision tree, and Random forest (RF). The authors reported that the Random forest archived the accuracy of 98.21% and had a detection rate of 98.50%.

3.2 Intrusion Detection System

An intrusion detection system (IDS) is a system that analyzes when the system has signs of possible incidents, violations, or imminent threats. Today, IDS have to deal with enormous volumes and high dimensional data, and with sophisticated and continuously changing behavior as well as the necessity of real-time protection. As a consequence, researchers have always been seeking new methods that provide data in a robust and efficient manner.

The natural features of SI make it ideal for intrusion detection and rule mining purposes. Indeed, SI approaches aim at solving difficult tasks by using multiple simple agents without the presence of a centralized authority. To be more precise, a difficult IDS problem will be split into multiple simple ones and assigned to the agents. Because of the advantages of SI algorithms, they could make the IDS more robust, adaptive, self-organized, and cost-efficient. Hence, various bio-inspired metaheuristic methods have utilized to improve the efficiency of IDSs in recent years.

The ACO approaches presented in Ref. [35] showed its benefits for feature selection for intrusion detection. The main contributions of their work showed that the ACO could be used for selecting robust features. Their work was evaluated on the KDD'99 Cup dataset and was able to reach an accuracy of 98.29%. Likewise, the authors in [3] utilized ACO algorithms for features selection and nearest neighbor classifier as a trained classifier for identifying various kinds of attack including Denial of Service (DoS), User to Root (U2R), Remote to Local (R2L), and Probing attacks. The algorithm was tested on the KDD'99 Cup, and NSL-KDD datasets were able to reach accuracy rates of 98.90% and the false alarm rate of 2.59%. In a later study, Botes et al. [11] presented a new method namely Ant Tree Miner (ATM) classified, which is a decision tree using ACO instead of conventional techniques such as C4.5 and CART [39], for intrusion detection. Using NSL-KDD datasets, their approach archived the accuracy of 65% and false alarm rate 0%.

Contemporary with ACO, PSO was also used and achieved satisfactory results when utilized in IDS. In [1], Aburomman et al. applied ensemble techniques to the intrusion detection problem. As base classifiers for these ensemble methods, they used support vector machine (SVM) and k-nearest-neighbor (kNN), on the KDD'99 dataset. After that, three new ensembles were created by using three techniques: PSO, another based on a variant of the PSO, where parameters were optimized by the local unimodal sampling (LUS), and lastly the Weighted Majority Algorithm (WMA). Their experiment showed that the ensemble technique was able to improve the results obtained from the base classifiers alone. Furthermore, the ensembles generated by PSO techniques archived better results than the traditional WMA.

In a later study [47], the authors presented an IDS using binary PSO and kNN. The proposed method consists of feature selection and classification step. Based on the results obtained, the algorithm showed excellent performance, and the proposed hybrid algorithm raised the accuracy generated by KNN by up to 2%. Meanwhile, Ali et al. [4] introduced a learning model for fast learning network (FLN) based on PSO named PSO-FLN, and the model had been utilized to the problem of IDS. The PSO-FLN model was tested on the KDD'99 Cup datasets and achieved the highest testing accuracy compared to other meta-heuristic algorithms.

In the recent study by Chen et al. [12], a multi-level adaptive coupled intrusion detection method combining white list technology and machine learning was presented. The white list was used to filter the communication, and the machine learning model was used to identify abnormal communication. In this article, the adaptive PSO algorithm and Artificial Fish Swarm (AFS) algorithm were used to optimize the parameters for the machine learning model. The method was tested on KDD99 Cup, Gas Pipeline, and industrial field datasets. The empirical result showed that the proposed model is efficient with various attack types.

In addition to ACO and PSO, other SI-based algorithms were also adopted for identifying malicious activities on the system. In [23], the authors introduced the Fuzzified Cuckoo based Clustering Technique for anomaly detection. The technique consists of two phases: the training phase and the detection phase. In the training phase, Cuckoo Search Optimization (CSO), K-means clustering, and Decision Tree Criterion (DTC) were combined to evaluate the distance functions. In the detection phase, a fuzzy decisive approach was utilized to identify the anomalies based on input data and previously computed distance functions. Experimental results showed that the model was effective with an accuracy rate of 97.77% and a false alarm rate of 1,297%.

Meanwhile, the authors in [26] incorporated Artificial Bee Colony and Artificial Fish Swarm algorithms to cope with the complex IDS problems. In this work, a hybrid classification method based on the ABC and AFS algorithms was proposed to improve the detection accuracy of IDS. The NSL-KDD and UNSW-NB15 datasets were used to evaluate the performance of these methods. Based on the results obtained, the proposed model was efficient with a low false alarm rate and high accuracy rate.

In later research, Garg et al. [24] proposed a hybrid model for network anomaly detection in cloud environments. The model utilized Grey Wolf Optimization (GWO) and Convolutional Neural Network (CNN) for feature extraction and identifying the anomalies on real-time network traffic streams. The empirical result showed that the proposed model is efficient with a low false alarm rate and high detection rate.

In Ref. [43], the authors adopted the firefly algorithm for feature selection and C4.5, Bayesian Networks classifier for detection network intrusion. The proposed approach was tested on the KDD'99 Cup dataset, and obtained a promising result and outperformed the compared method for feature selection.

Recently, research conducted by Gu et al. [25] introduced an IDS based on SVM with the Tabu-Artificial Bee Colony for feature selection and parameter optimization simultaneously. The main contributions of their work included the adopting of Tabu Search algorithm to improve the neighborhood search of ABC so that it could speed up the convergence and prevented stuck in the local optimum. According to their experiments, although the accuracy rate was high 94.53%, the false alarm rate was 7.03%.

3.3 Other Domains

In addition to conventional threats, other cybersecurity issues and solutions utilizing SI based approach are equally worthy of attention.

3.3.1 Spam detection

The authors in [33], designed a spam categorization technique using a modified cuckoo search to enhance the spam classification. In their work, the step size-cuckoo search was utilized for feature extraction, and the SVM was used for classification. The proposed approach was tested on two spam datasets: Bare-ling, Lemm-ling, and obtained a competitive result.

Later, research conducted by [44] proposed a system to filter the spam message of Facebook using SI-based and machine learning techniques. The PSO algorithm was adopted for feature selection and the SVM, decision tree for classification. The authors claimed that the proposed system was efficient. Unfortunately, the details of the results were not provided.

Recently, Aswani et al. [7] provided a hybrid approach for detecting the spam profiles on Twitter using social media analytics and bio-inspired computing. Specifically, they utilized a modified K-Means integrated Levy flight Firefly Algorithm (LFA) with chaotic maps to identify spammers. A total of 14,235 profiles were used to evaluate the performance of the method. The empirical result showed that the proposed model was efficient with an accuracy of 97.98%.

3.3.2 Phishing detection

Tayal and Ravi [49] developed an algorithm called Particle Swarm Optimization trained Classification Association Rule Mining for classifying phishing email and phishing websites. In this work, the binary PSO was adopted to generate the association rules from the transactional database. Based on the results obtained, the algorithm obtained an accuracy of 83% and 88% for phishing email detection and phishing URL detection, respectively.

In a later study, the authors in [13] presented a model for detecting phishing websites based on the improved BP neural network and dual feature evaluation mechanism named DF.GWO-BPNN. In this model, the Grey Wolf Algorithm was adopted to optimize the parameters for the BP neural network. The dataset used to evaluate the model was collected from Phishtank, the International Anti-Phishing Organization (APWG), and Security Alliance. The authors reported that their model achieved an accuracy of 98.78%.

3.3.3 Designing virus prototypes for research

Regarding the advancing in SI and technologies, it is logical to expect that, shortly, SI-based malware will appear. To be ready for future and unknown SI-based malware threats, the authors in [54] outlined a hypothetical swarm malware as a background for a future antimalware system. More precisely, the swarm virus prototype simulated a swarm system behavior, and its information was stored and visualized in the form of a complex network. As a further improvement, the authors in [16] fused swarm base intelligence, neural network, and a classical computer virus to form a neural swarm virus. This virus prototype utilized the NN as a the "trigger conditions" to propagate. Technically, this method will be an extreme challenge for malware analysts to figure out what category of target the malware was looking for, or what is the specific target to trigger conditions.

4. Discussion

Over the last few years, researchers have shown a great deal of interest in adopting bio-inspired mechanisms in general and swarm intelligence in particular to issues in terms of cybersecurity. Tables 1 and 2 summarize some recent studies that are adopting SI-based techniques for cybersecurity issues, the details of which were described earlier.

Table 1 summarizes the swarm-based methods for fighting malware. This table consists of the following entries: name of the SI technique adopted, and the main ideas utilized in order to guide the search. From the table, one can notice the PSO algorithm is widely utilized in the domain of malware detection, especially on the Android environment.

Furthermore, most of the existing work utilizes an SI-based approach to optimize the parameters or features. Additionally, the swarm-based techniques were incorporated with other machine learning techniques to enhance the overall performance of the models. What's more, most of the existing methods obtained encouraging results, which proved that an SI-based technique can be adopted for identifying new malware.

Table 1. Summary of SI-based techniques for malware detection and prevention

Ref.	*Year*	*SI tech.*	*Main idea*	*Result*
[2]	2016	PSO	Combining adaptive neural fuzzy interface system and PSO to optimize the parameters	RMSE: 0.43106 $R2 : 0.7721$
[45]	2016	PSO	Integrating PSO and GA with Random forest algorithm	Accuracy: 88.4%
[6]	2017	PSO	IG and PC for ranking the permissions and API calls and PSO to optimize the parameters	Accuracy: 89%
[5]	2018	PSO	Adopting PSO for feature selection and RF to identify obfuscated Windows malware	Accuracy: 99.6%
[46]	2018	WOH	Applying WOA to extract features Nave Bayes, J48, and RF for classification	Accuracy: 98.21%
[9]	2019	PSO	Improving PSO algorithm for feature selection through rough set of permissions	Accuracy: 97.86%

Table 2 epitomizes the swarm-based approaches for intrusion detection. This table includes the name of the SI algorithm, the diversity mechanism, data set used for validating, and the performance metrics. From the summary it is obvious that most of the SI systems have given excellent results. This confirms that the SI-based systems are robust, highly adaptive, and capable of detecting novel attack types.

One point to be taken into account is that the PSO algorithm is rarely used as the exclusive method for classification. The main reason is that PSO can become stuck in local solutions in the early stages of its execution and can be easily outperformed.

For the datasets part, KDD'99 and NSL-KDD was used by most of the IDS to evaluate the performance. This may not reflect the real world situations.

Furthermore, the majority of the IDS utilized SI algorithms for optimizing parameters and features. It is clear that the majority of the relative research treats this technique as a supplementary step to some other machine learning classifier which conducts the main part of the classification. Fewer proposals as shown in [11, 26], on the other hand, proposed to use swarm-based method for enhancing the classification process.

Although SI approaches have obtained encouraging achievement, there are still avenues for further investigation, as presented below:

- We are still lacking research about adopting SI technique for detecting malware on the Windows environment. Therefore research on this area is urgent.

Table 2. Summary of SI-based techniques for intrusion detection

Ref.	Year	SI tech.	Diversity technique	Dataset	Result
[35]	2016	ACO	ACO for feature selection SVM for building detection model	KDD'99	Accuracy: 98.29% FAR: 2%
[3]	2016	ACO	ACO for feature selection Nearest neighbor as a trained classifier	KDD'99, and NSL-KDD	Accuracy: 98.90% FAR: 2.59%
[1]	2016	PSO	Ensemble techniques with SVM and kNN PSO for creating ensemble of classifiers	KDD'99	Accuracy: ≈ 92.79% FAR: Not Available
[11]	2017	ACO	Introduced Ant Tree Miner classified, which adopted ACO to build a decision tree	NSL-KDD	Accuracy: 65.0% FAR: 0%
[47]	2017	PSO	PSO for feature selection kNN for classification	KDD'99	Accuracy: ≈ 98.64% FAR: Not Available
[4]	2018	PSO	Presented a learning model for fast learning network based on PSO	KDD'99	Accuracy: 99.64% FAR: very low
[23]	2018	CSO	Proposed Fuzzied Cuckoo based Clustering Technique for anomaly detection	UCI-ML, and NSL-KDD	Accuracy: 97.77% FAR: 1.297%
[26]	2018	ABC + AFS	Incorporated ABC and AFS for classification	NSL-KDD, and UNSW-NB15	Accuracy: 99% FAR: 0.01%
[12]	2019	PSO + AFS	Adoptive PSO + AFS for optimizing the parameters	KDD99, Gas Pipeline, and industrial field	DR: highest 95%, lowest: 65.5% FAR: highest 8.1%, lowest: 0.02%

(Contd.)

[24]	2019	GWO	GWO for feature extraction CNN for identifying the anomalies	DARPA98, and KDD99	Accuracy: 97.92%, 98.42% FAR: 3.60%, 2.22%
[43]	2019	FA	FA for feature selection C4.5, Bayesian Networks for detection	KDD'99	DoS: Acc.: 99.98%, FAR: 0.01% Probe: Acc.: 93.42%, FAR: 0.01% R2L: Acc.: 98.73%, FAR: 0% U2R: Acc.: 68.97%, FAR: 0%
[25]	2019	ABC	Adopted TS algorithm to improve ABC SVM for classification	KDD'99	Accuracy: 94.53% FAR: 7.03%

- Most of the existing literature about IDS lack s the use of real data. So this is another aspect that needs to be considered.
- Adopting SI method for the cybersecurity issues of cloud computing systems is an interesting field of research.
- The combination of multi SI algorithms to solve cybersecurity problems is also a potential for investigation.

5. Conclusion

Dramatic advances in information technology have led to the emergence of new challenges for cybersecurity. The computational complexity of cyber-attacks require new approaches which are more robust, scale-able, and flexible. This chapter focuses on the application of the SI-based techniques in cybersecurity issues. Specifically, the application of SI in malware detection, intrusion detection and others domains such as spam detection, phishing detection. Our review shows that there is clear motivation to further investigate SI techniques for complex cybersecurity problems in the future.

Acknowledgements

The following grants are acknowledged for the financial support provided for this research: Grant of SGS No. SP2019/137, VSB Technical University of Ostrava.

References

1. A.A. Aburomman and M.B.I. Reaz. A novel svm-knn-pso ensemble method for intrusion detection system. Applied Soft Computing 38, 360–372 (2016).
2. F. Afifi, N.B. Anuar, S. Shamshirband and K.K.R. ChooDyhap. Dynamic hybrid anfis-PSO approach for predicting mobile malware. PloS One 11(9), e0162627 (2016).
3. M.H. Aghdam and P. Kabiri. Feature selection for intrusion detection system using ant colony optimization. IJ Network Security 18(3), 420–432 (2016).
4. M.H. Ali, B.A.D. Al Mohammed, A. Ismail and M.F. Zolkipli. A new intrusion detection system based on fast learning network and particle swarm optimization. IEEE Access 6, 20255–20261 (2018)
5. Z. Ali and T.R. Soomro. An efficient mining based approach using pso selection technique for analysis and detection of obfuscated malware. Journal of Information Assurance & Cyber Security 2018, 1–13 (2018).
6. A. Altaher and O.M. Barukab. Intelligent hybrid approach for android malware detection based on permissions and API calls. International Journal of Advanced Computer Science and Applications 8(6), 60–67 (2017).

7. R. Aswani, A.K. Kar and P.V. Ilavarasan. Detection of spammers in twitter marketing: a hybrid approach using social media analytics and bio inspired computing. Information Systems Frontiers 20(3), 515–530 (2018).

8. G. Beni and J. Wang. Swarm intelligence in cellular robotics systems. Proceedings of NATO Advanced Workshop on Robots and Biological System, pp. 703–712 (1989).

9. A. Bhattacharya, R.T. Goswami and K. Mukherjee. A feature selection technique based on rough set and improvised PSO algorithm (PSORS-FS) for permission based detection of android malwares. International Journal of Machine Learning and Cybernetics 10(7), 1893–1907 (2019).

10. E. Bonabeau, D.d.R.D.F. Marco, M. Dorigo and G. Theraulaz. Swarm intelligence: from natural to artificial systems. No. 1, Oxford University Press (1999).

11. F.H. Botes, L. Leenen and R. De La Harpe. Ant colony induced decision trees for intrusion detection. 16th European Conference on Cyber Warfare and Security, pp. 53–62 (2017).

12. W. Chen, T. Liu, Y. Tang and D. Xu. Multi-level adaptive coupled method for industrial control networks safety based on machine learning. Safety Science 120, 268–275 (2019).

13. W. Chen, X.A. Wang, W. Zhang and C. Xu. Phishing detection research based on PSO-BP neural network. International Conference on Emerging Internetworking, Data & Web Technologies, pp. 990–998. Springer (2018).

14. B. Christian and M. Daniel. Swarm Intelligence Introduction and Application. Natural Computing Series, Springer (2008).

15. S.C. Chu, P.W. Tsai, and J.S. Pan. Cat swarm optimization. Pacific Rim International Conference on artificial intelligence, pp. 854–858. Springer (2006).

16. T.T. Cong, I. Zelinka and R. Senkerikr. Neural swarm virus. Proceedings of 7th Joint International Conferences on Swarm, Evolutionary and Memetic Computing Conference (SEMCCO 2019) & Fuzzy And Neural Computing Conference (FANCCO 2019) (2019).

17. D. Davendra and I. Zelinka. Self-organizing migrating algorithm. New Optimization Techniques in Engineering (2016).

18. M. Dorigo, E. Bonabeau and G. Theraulaz. Ant algorithms and stigmergy. Future Generation Computer Systems 16(8), 851–871 (2000).

19. M. Dorigo, M.Dorigo, V. Manjezzo and M. Trubian. Ant system for job-shop scheduling. Belgian Journal of Operations Research 34, 39–53 (1994).

20. M. Dorigo, V. Maniezzo and A. Colorni. Positive Feedback as a Search Strategy. (1991).

21. R. Eberhart and J. Kennedy. A new optimizer using particle swarm theory. *In*: MHS '95. Proceedings of the Sixth International Symposium on Micro Machine and Human Science, pp. 39–43. IEEE (1995).

22. A.P. Engelbrecht. Computational Intelligence: An Introduction. John Wiley & Sons (2007).

23. S. Garg and S. Batra. Fuzzified cuckoo based clustering technique for network anomaly detection. Computers & Electrical Engineering 71, 798–817 (2018).

24. S. Garg, K. Kaur, N. Kumar, G. Kaddoum, A.Y. Zomaya and R. Ranjan. A hybrid deep learning based model for anomaly detection in cloud data centre networks. IEEE Transactions on Network and Service Management (2019).

25. T. Gu, H. Chen, L. Chang and L. Li. Intrusion detection system based on improved ABC algorithm with tabu search. IEEJ Transactions on Electrical and Electronic Engineering (2019).

26. V. Hajisalem, and S. Babaie. A hybrid intrusion detection system based on ABC-AFS algorithm for misuse and anomaly detection. Computer Networks 136, 37–50 (2018).

27. F. Heppner and U. Grenander. A stochastic nonlinear model for coordinated bird flocks. The ubiquity of Chaos 233, 238 (1990).

28. D. Karaboga. An idea based on honey bee swarm for numerical optimization. Tech. Rep., Technical report-tr06, Erciyes University, Engineering Faculty (2005).

29. D. Karaboga and B. Basturk. A powerful and efficient algorithm for numerical function optimization: artificial bee colony (ABC) algorithm. Journal of Global Optimization 39(3), 459–471 (2007).

30. J. Kennedy and R. Eberhart. Particle swarm optimization (PSO). *In*: Proc. IEEE International Conference on Neural Networks, Perth, Australia, pp. 1942–1948 (1995).

31. J. Kennedy. Particle swarm optimization. Encyclopedia of Machine Learning, pp. 760–766 (2010).

32. K. Krishnanand and D. Ghose. Glowworm swarm optimisation: a new method for optimising multi-modal functions. International Journal of Computational Intelligence Studies 1(1), 93–119 (2009).

33. T. Kumaresan and C. Palanisamy. E-mail spam classification using s-cuckoo search and support vector machine. International Journal of Bio-Inspired Computation 9(3), 142–156 (2017).

34. Li, X.I. An optimizing method based on autonomous animats: fish-swarm algorithm. Systems Engineering – Theory & Practice 22(11), 32–38 (2002).

35. T. Mehmod and H.B.M. Rais. Ant colony optimization and feature selection for intrusion detection. Advances in Machine Learning and Signal Processing, pp. 305–312. Springer (2016).

36. S. Mirjalili and A. Lewis. The whale optimization algorithm. Advances in Engineering Software 95, 51–67 (2016).

37. S. Mirjalili, S.M. Mirjalili and A. Lewis: Grey wolf optimizer. Advances in Engineering Software 69, 46–61 (2014).

38. S. Olariu and A.Y. Zomaya. Handbook of Bioinspired Algorithms and Applications. Chapman and Hall, CRC (2005).

39. F.E. Otero, A.A. Freitas and C.G. Johnson. Inducing decision trees with an ant colony optimization algorithm. Applied Soft Computing 12(11), 3615–3626 (2012).

40. B. Panigrahi, Y. Shi and M. Lim. Handbook of Swarm Intelligence. Series: Adaptation, Learning, and Optimization (2011).

41. K.M. Passino. Biomimicry of bacterial foraging for distributed optimization and control. IEEE Control Systems Magazine 22(3), 52–67 (2002).

42. C.W. Reynolds. Flocks, Herds and Schools: A Distributed Behavioral Model, vol. 21. ACM (1987).

43. B. Selvakumar and K. Muneeswaran. Firefly algorithm based feature selection for network intrusion detection. Computers & Security 81, 148–155 (2019).

44. M.K. Sohrabi and F. Karimi: A feature selection approach to detect spam in the facebook social network. Arabian Journal for Science and Engineering 43(2), 949–958 (2018).

45. M. Sujithra and G. Padmavathi. Enhanced permission based malware detection in mobile devices using optimized random forest classifier with PSO-GA. Research Journal of Applied Sciences, Engineering and Technology 12(7), 732–741 (2016).

46. S.A. Sulaiman, O.S. Adebayo, I. Idris and S.A. Bashir. Android malware classification using whale optimization algorithm. i-manager's Journal on Mobile Applications and Technologies 5(2), 37 (2018).

47. A.R. Syarif and W. Gata. Intrusion detection system using hybrid binary PSO and K-nearest neighborhood algorithm. *In*: 2017 11th International Conference on Information & Communication Technology and System (ICTS), pp. 181–186. IEEE (2017).

48. P. Szor. The Art of Computer Virus Research and Defense. Pearson Education (2005).

49. K. Tayal and V. Ravi. Particle swarm optimization trained class association rule mining: application to phishing detection. *In*: Proceedings of the International Conference on Informatics and Analytics, p. 13. ACM (2016).

50. D. Ucci, L. Aniello and R. Baldoni. Survey of machine learning techniques for malware analysis. Computers & Security (2018).

51. X.S. Yang. Firefly algorithms for multimodal optimization. *In*: International Symposium on Stochastic Algorithms, pp. 169–178. Springer (2009).

52. X.S. Yang. A new metaheuristic BAT-inspired algorithm. *In*: Nature Inspired Co-operative Strategies for Optimization (NICSO 2010), pp. 65–74. Springer (2010).

53. I. Zelinka. Somaself-organizing migrating algorithm. *In*: New Optimization Techniques in Engineering, pp. 167–217. Springer (2004).

54. I. Zelinka, S. Das, L. Sikora and R. Šenke□ik. Swarm virus-next-generation virus and antivirus paradigm? Swarm and Evolutionary Computation 43, 207–224 (2018).

6

Emergence of Complex Phenomena in a Simple Reversible Cellular Space

Kenichi Morita

Hiroshima University, Higashi-Hiroshima, 739-8527, Japan
Email: km@hiroshima-u.ac.jp

1. Introduction

A cellular automaton (CA) is a system consisting of a large number of identical finite state automata, called cells, interconnected uniformly in a space. Hence it is considered a kind of parallel processing model. It is often used to study how complex phenomena emerge from a collection of primitive cells. A typical example is the Game-of-Life CA [2, 4, 5]. On the other hand, a CA can also be regarded to be an abstract model of a physical space. In particular, a reversible cellular automaton (RCA) is a spatiotemporal model that reflects physical reversibility, one of the fundamental laws of nature. An RCA is a CA whose global function (i.e., the mapping that determines the transition of the configurations of the cellular space) is injective. It is thus suited for studying how high-order functions appear from a collection of simple reversible elements.

Here, we use the framework of an *elementary triangular partitioned cellular automaton* (ETPCA) [10, 11]. A cell of an ETPCA is triangular, and divided into three parts, each of which has only two states. The next state of a cell is determined by the three adjacent parts of its neighbour cells. Figure 1 shows its cellular space and how the state of a cell changes. It has been shown that, in a partitioned CA, its global function is injective if and only if its local function (i.e., the mapping that determines the transition of the states of each cell) is injective [12, 10]. Hence, a partitioned CA makes it easier to design an RCA. In addition, in a reversible partitioned CA, each cell itself becomes a reversible finite state machine.

ETPCAs are very simple, since each of their local functions are described by only four local transition rules. Here, we focus on a specific

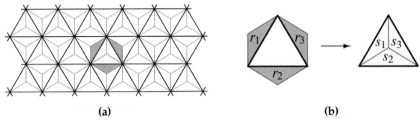

Fig. 1. A triangular partitioned cellular automaton. (a) Its cellular space.
(b) A local transition rule.

reversible ETPCA 0347, where 0347 is its identification number in the
class of ETPCAs. Figure 2 gives us its local transition rules, which define
its local function. Note that this local function is injective, since there
is no pair of local transition rules that have the same right-hand sides.
ETPCA 0347 exhibits complex behaviour as in the case of the Game-of-
Life CA. There are various interesting objects (patterns) in it. In particular,
there exists a glider, a space-moving pattern. Interactions of a glider with
another glider or other patterns show a fascinating evolution processes
[11]. It was shown that a Fredkin gate, a universal reversible gate, is
embeddable in ETPCA 0347. In addition, reversible Turing machines can
be simulated in it [8].

In this survey, we discuss how we can find a pathway of constructing
universal reversible computers from a large number of primitive cells that
obey the reversible local function given in Fig. 2. We first identify several
basic patterns in ETPCA 0347. Secondly, we give useful fundamental
phenomena caused by interacting basic patterns. Third, a reversible
logic element suitable for constructing reversible computers is supposed,
and it is implemented in ETPCA 0347 by utilising the fundamental
phenomena. Finally, a reversible computer is assembled using these logic
elements, and realised in the cellular space. In this way, we can compose
any reversible computer from a large number of simple reversible cells in
a hierarchical and systematic way.

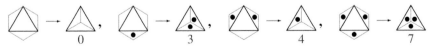

Fig. 2. Local function of the reversible ETPCA 0347 defined
by four local transition rules.

2. Elementary Triangular Partitioned Cellular Automata

A *partitioned cellular automaton* (PCA) is a subclass of a standard CA, where
a cell is divided into several parts, and each part has a state set. Figure
1(a) shows the cellular space of a two-dimensional three-neighbour
triangular PCA (PCA). In a triangular PCA, the next state of a cell is
determined by the states of the adjacent parts of the three neighbor

cells, not by the states of the whole three cells. Namely, the next state is determined by a set of local transition rules of the form shown in Fig. 1 (b). Thus the set of local transition rules defines a *local function* of the PCA. A *configuration* of a PCA is a state of the whole (infinite) cellular space of it. Applying the local function to all the cells in parallel, a *global function*, which gives a transition relation among configuration, is obtained.

We say a PCA is *locally reversible* if its local function is injective, and *globally reversible* if its global function is injective. It is easy to test local reversibility of a given PCA by checking if there is no pair of local transition rules that have the same right-hand sides. It has been shown that global reversibility and local reversibility are equivalent (Lemma 1). Hence such a PCA is simply called a *reversible PCA* (RPCA).

Lemma 1 [12] *A PCA A is globally reversible if and only if it is locally reversible.*

By this lemma, to obtain a reversible CA, it is sufficient to give a locally reversible PCA. Thus, the framework of PCAs makes it easy to design reversible CAs.

A triangular PCA is called *isotropic* (or *rotation-symmetric*), if, for each local transition rule, the rules obtained by rotating both sides of it by a multiple of 60° exist. A triangular PCA is called an *elementary triangular PCA* (ETPCA) if it is isotropic and each part of the cell has the state set {0, 1}. In the following figures, the states 0 and 1 are indicated by a blank and a particle •, respectively. ETPCAs are one of the simplest classes of two-dimensional PCAs. Yet, this class still contains many interesting PCAs as in the case of one-dimensional elementary CAs (ECAs) [16, 17].

In ETPCAs, there are two kinds of cells, i.e., an up-triangle cell and a down-triangle cell, as in Fig. 3. But, they differ only on their directions, and, of course, they have the same local function. Here, we denote their states shown in Fig. 3 by a triplet $(s_1, s_2, s_3) \in \{0, 1\}^3$.

Since an ETPCA is isotropic, and each part of a cell has two states, its local function is defined by only four local transition rules. Hence, an ETPCA can be specified by a four-digit number $wxyz$ where $w, z \in \{0, 7\}$ and $x, y \in \{0, 1, \ldots, 7\}$ as shown in Fig. 4. Thus, there are 256 ETPCAs. Here, w and z must be 0 or 7, because ETPCAs are isotropic and deterministic. The ETPCA with the identification number $wxyz$ is

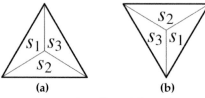

(a)   (b)

Fig. 3. (a) An up-triangle cell, and (b) a down-triangle cell in the space of ETPCA, whose states are $(s_1, s_2, s_3) \in \{0, 1\}3$.

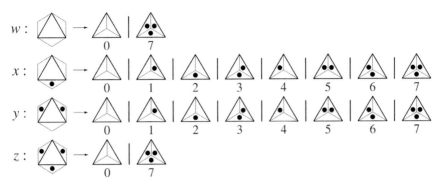

Fig. 4. Representing an ETPCA by a four-digit number $wxyz$, where $w, z \in \{0, 7\}$ and $x, y \in \{0, 1, \ldots, 7\}$. The states 0 and 1 are represented by a blank and •, respectively. Vertical bars indicate alternatives of the right-hand side of each local transition rule.

denoted by ETPCA $wxyz$. Figure 2 shows the local function of the ETPCA 0347.

By Lemma 1, it is easy to see the following: An ETPCA $wxyz$ is reversible if and only if

$$(w, z) \in \{(0, 7), (7, 0)\} \wedge$$
$$(x, y) \in \{1, 2, 4\} \times \{3, 5, 6\} \cup \{3, 5, 6\} \times \{1, 2, 4\}.$$

Hence, ETPCA 0347 is reversible, and there are 36 reversible ones in total.

For an ETPCA $0xyz$, we define a *quiescent state* as the state $(0, 0, 0)$. In such an ETPCA, if all the neighbour cells are quiescent states, then the centre cell becomes quiescent at the next time step. A *finite configuration* is one such that all but finite number of cells are quiescent. An *infinite configuration* is one such that infinitely many cells are non-quiescent.

3. A Pathway from a Reversible Microscopic Law to Reversible Computers

In this section, we explain how reversible computers can be realised in the simple reversible ETPCA No. 0347 based on the results in [8, 11]. It is done from bottom to top in a hierarchical way. In Sec. 3.1, basic patterns (or objects) in ETPCA 0347, which are used in the rest of this article, are identified. In Sec. 3.2, several useful fundamental phenomena resulted from interactions of basic patterns are shown. In Sec. 3.3, a reversible logic element with one-bit memory (RLEM) for constructing reversible Turing machines is explained, and realisation method of a specific RLEM No.4-31 by utilising fundamental phenomena is given. In Sec. 3.4, how reversible Turing machines are constructed out of RLEM 4-31 is explained. Based on this, configurations that simulate reversible Turing machines are shown.

Note that, generally, it is not easy to follow an evolving processes of configurations of ETPCA 0347 by hand, though the local function itself is quite simple. So, we made a program for simulating them. The slide file in [11] contains simulation movies of various examples. Furthermore, we created an emulator of ETPCA 0347 on a general purpose CA simulator *Golly* [15]. The file of the emulator is found in [9].

3.1 Basic Patterns in ETPCA 0347

A *pattern* is a finite segment of a configuration. In this section, we give several useful patterns in ETPCA 0347, and watch how they behave. They can be considered to be elementary objects for composing more complex objects. Placing such patterns appropriately in the cellular space we can construct configurations that perform interesting tasks.

Patterns are classified into three categories. They are a periodic pattern, a space-moving pattern, and an expanding pattern [11]. As we shall see, some of periodic patterns and a space-moving pattern are very useful. They are a *block*, a *fin*, a *rotator*, and a *glider*, which are called *basic patterns*.

3.1.1 Periodic patterns

A *periodic pattern* (or a *pattern of period p*) is one that satisfies the following condition: Starting from the configuration consisting of one copy of it, the same pattern appears at the same position after p time steps $(p > 0)$. A *stable pattern* is a pattern of period 1, which is a special case of a periodic pattern. It should be noted that in ETPCA 0347 there is no "eventually periodic pattern" (i.e., a pattern that becomes periodic after one or more transient steps), since ETPCA 0347 is reversible. There are many small periodic patterns in ETPCA 0347 [11]. Among them, the following three are important.

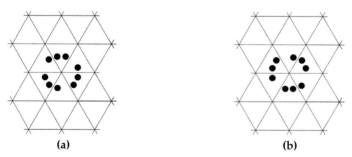

(a)   (b)

Fig. 5. Blocks of (a) type I and (b) type II in ETPCA 0347.
They are stable patterns.

A *block* is a stable pattern shown in in Fig. 5. There are two kinds of blocks, i.e., type I (Fig. 5(a)) and type II (Fig. 5(b)). They will be used to control the direction of a space-moving pattern called a glider (Sec. 3.2.2).

A *fin* is a periodic pattern shown in Fig. 6. It rotates clockwise with a period 6. As we shall see below, a fin appears when a glider collides a block (Sec. 3.2.2). It is also used as a kind of memory, since its position can be shifted by colliding a glider (Sec. 3.2.3).

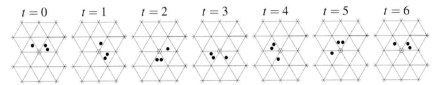

Fig. 6. A fin is a pattern of period 6 in ETPCA 0346. It rotates around the point ○ clockwise.

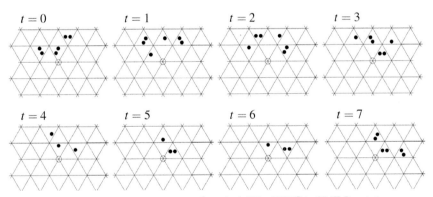

Fig. 7. A rotator is a pattern of period 42 in ETPCA 0347. It rotates around the point ○.

A *rotator* is a periodic pattern of period 42 shown in Fig. 7. Like a fin, it rotates around some point. If a glider collides a block, it will be decomposed into a fin and a rotator (Sec. 3.2.2).

3.1.2 Space-moving pattern

A *space-moving pattern* (or a *spaceship*) is one such that after some time steps p ($p > 0$) the same pattern (not rotated one) appears at a different position. Thus, it moves straight ahead in the cellular space of ETPCA 0347 if no obstacle exists. The integer p is also called the *period* of the space-moving pattern.

The pattern shown in Fig. 8 is a space-moving pattern. It is called a *glider* as in the case of the Game-of-Life CA. It swims in the cellular space like a fish or an eel. It travels a unit distance, the side-length of a triangle,

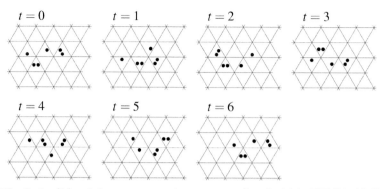

Fig. 8. A glider. It is a space-moving pattern of period 6 in ETPCA 0347.

in 6 steps. Thus, its speed is 1/6. By rotating it appropriately, it can move in any of the six directions.

A glider shows fascinating behaviour when it interacts with blocks (Sec. 3.2.2), a fin (Sec. 3.2.3), and another glider (Sec. 3.2.4). It will be used as a signal when we construct logic circuits in the cellular space.

So far, it is unknown whether there is another space-moving pattern that is not composed of two or more glider patterns.

3.1.3 Expanding patterns

An *eventually expanding pattern* (or simply *expanding pattern*) is one such that the diameter of the pattern grows indefinitely as it evolves (though we do not give a definition of "diameter" here, it should be defined appropriately). Namely, for any integer $d_0 > 0$, there exists an integer $t_0 > 0$ such that the diameter of the pattern at time t_0 is greater than or equal to d_0. It should be noted that at first its diameter may decrease, but eventually it grows bigger and bigger.

A pattern consisting of one particle is an example of an expanding pattern. If we start from it, a disordered pattern and many gliders are generated, and the whole pattern grows indefinitely (Fig. 9).

Any expanding pattern also expands to the negative time direction. Namely, if a pattern P at $t = 0$ is an expanding pattern, then for any $d_0 > 0$, there is a pattern P_t at $t_0 < 0$ that becomes P at $t = 0$, and its diameter is greater than or equal to d_0. This fact is explained as follows. First, we observe that for each pattern P_t at time t, the previous pattern P_{t-1} at $t - 1$ is easily obtained by reversely applying the local transition rules given in Fig. 2. Now, assume, on the contrary, there is an integer d_{max} such that the diameter of the pattern P_t at time t is less than d_{max} for all $t < 0$. Then, the total number of different patterns that appear at $t < 0$ is finite. Hence, there are two integers $t_1 < t_2 < 0$ for which the patterns P_{t_1} and P_{t_2} are the same. Therefore, P_{t_1} is either a periodic pattern or a space-moving pattern, and not an expanding pattern.

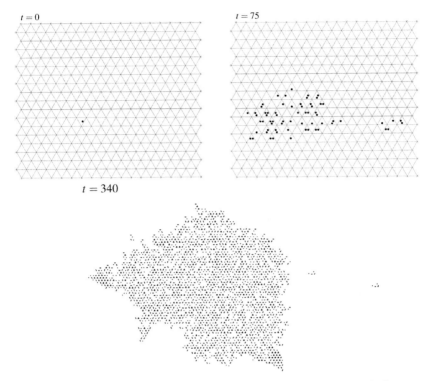

Fig. 9. Evolution process starting from the one-particle pattern ($t = 0$) [11]. It is an expanding pattern, since the sizes of the patterns grow indefinitely as it evolves.

In the reversible ETPCA 0347, a disordered pattern like the one in Fig. 9 ($t = 340$) appears very often, even if we compose a configuration out of periodic patterns and gliders. Therefore, when we want to give a configuration that performs some intended task, it should be designed so that it never generates a disordered pattern.

There are many other expanding patterns. A pattern consisting of two gliders that move in opposite directions is a simple example. In [11] it is shown that glider guns, which generate gliders periodically, exist in ETPCA 0347. They are other examples of expanding patterns.

3.2 Fundamental Phenomena Resulted from Interactions of Basic Patterns

In this section, we observe various phenomena caused by interacting with the four kinds of basic patterns [11]. There are, of course, useful phenomena and (seemingly) useless ones. Here, we identify several useful phenomena as fundamental ones. They will be used in the following sections to construct reversible computers.

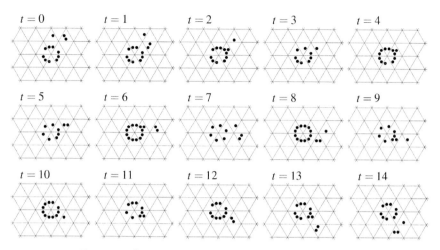

Fig. 10. A fin can travel around a block clockwise [11].

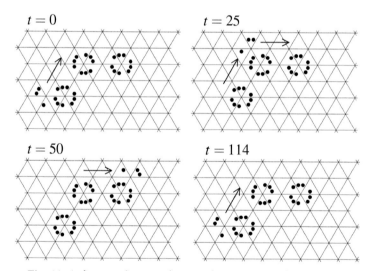

Fig. 11. A fin can also travel around a sequence of blocks [11].

3.2.1 Interaction of a fin and blocks

A fin can go around a block as in Fig. 10. We can see the whole pattern consisting of a block and a fin is a pattern of period 42. Furthermore, a fin can travel around a sequence of blocks as in Fig. 11. In this case the whole pattern is of period 114. Note that the sequence of blocks need not be placed on a straight line, i.e., it can be bent slightly. Also note that, since the number of blocks in the sequence is arbitrary, we can have a pattern with a longer period.

3.2.2 Interaction of a glider and blocks

Next, we observe various phenomena caused by colliding a glider with a sequence of blocks. We shall see that the moving direction of a glider can be controlled by appropriately arranging the sequence of blocks. In particular, backward-turn, right-turn, and U-turn of a glider are possible. Thus, such sequences of blocks are used as *turn modules* hereafter.

A *backward-turn module* is obtained by putting an appropriate type of a single block on a travelling path of a glider. We place a glider that moves eastward, and a block of type I as shown in Fig. 12 ($t = 0$). At $t = 12$ they collide. Then, the glider is split into a rotator and a fin. The rotator is, so to say, a "body" of the glider. Since the body has no fin, it cannot swim, and thus begins to rotate around the point indicated by ○ ($t = 16$). The fin travels around the block ($t = 38$) as in the case of Fig. 10. When the fin comes back to the original position, it interacts with the body ($t = 50$). By this, the rotation centre of the body is shifted upward by two cells, and the fin travels around the block once more ($t = 61$). At $t = 94$, the body and the fin meet again. By this, the fin is attached to the body, and a glider is reconstructed. Finally, the glider goes westward ($t = 97$). By above, backward-turn of the glider is realised. Note that, in [11],

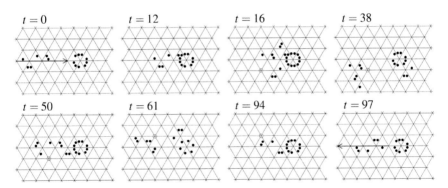

Fig. 12. Colliding a glider with a type I block [11]. It is used as a backward-turn module.

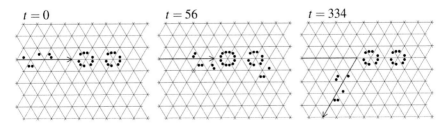

Fig. 13. Colliding a glider with a sequence of two blocks [11]. It is used as a 120°-right-turn module.

it is shown that if the block of type I is replaced by that of type II in Fig. 12, then the block will be broken. Hence, the type of a block should be chosen appropriately.

A *right-turn module* of a glider by 120° is realised by a sequence of two blocks (Fig. 13). Similar to the case of one block, the glider is first split into a rotator and a fin (t = 56). The fin travels around the blocks three times without interacting with the rotator. At the end of the fourth round, they meet to reconstruct a glider. Then, it goes to the south-west direction (t = 334). Hence, two blocks act as a 120°-right-turn module. Figure 14 shows that sequences of three blocks also acts as 120°-right-turn module. It has a shorter delay than the case of two blocks. Note that the three blocks need not be placed linearly.

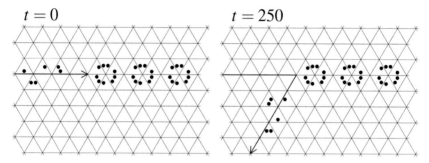

$t = 0$ $t = 250$

Fig. 14. Colliding a glider with a sequence of three blocks [11]. It also works as a 120°-right-turn module.

A *U-turn module* is given in Fig. 15. Also in this case, the glider is first split into a rotator and a fin (t = 36). But, slightly before the fin comes back to the start position, it meets the rotator, and a glider is reconstructed, which moves westward (t = 113). Here, the output path is different from the input path, while in the backward-turn module they are the same (Fig. 12).

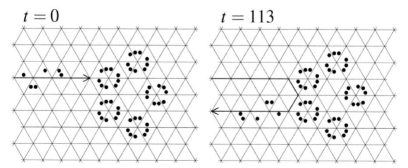

$t = 0$ $t = 113$

Fig. 15. Colliding a glider with a pattern consisting of five blocks [11]. It is used as a U-turn module.

In [11], a left-turn module consisting of 12 blocks is given. However, left-turn by 120° is realised by a two successive 120°-right-turns. On the other hand, right-turn by 60° is implemented by combining 120°-right-turn and backward-turn. It is also shown that the timing (or delay) of a glider, including its phase, can also be adjusted by these turn modules [11]. Thus the moving direction and the timing of a glider can be completely controlled.

3.2.3. Interaction of a glider and a fin

By interacting a glider with a fin, we obtain useful phenomena. If we collide a glider with a fin as in Fig. 16(a), the fin is pulled (it is also shifted southward by one cell), and the glider turns backward. Likewise, in Fig. 16(b), the fin is pushed, and the glider turns backward. Namely, by these operations, the position of a fin can be changed. Hence a fin can be used as a kind of memory. In Sec. 3.3.2, a fin is used to keep the state of a reversible logic element with one-bit memory.

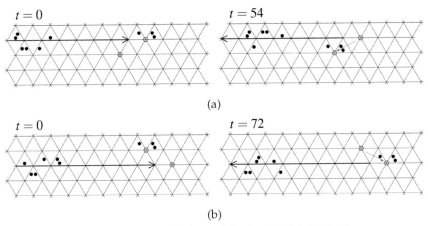

(a)

(b)

Fig. 16. Shifting a fin by a glider in ETPCA 0347 [8]
(a) Pulling, and (b) pushing.

Note that the phase of the fin does not change in these processes, i.e., the phase always becomes 0 if the time is a multiple of 6 as seen in Fig. 16. Therefore, controlling the process of successive shifts becomes easy.

3.2.4 Interaction of gliders

Here we consider the cases where two or more gliders interact, though we do not use these phenomena in this paper. Figure 17 shows that a kind of reversible logic gate called a *switch gate* is implemented by appropriately colliding two gliders. A switch gate is a 2-input 3-output

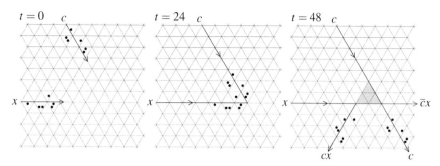

Fig. 17. A switch gate $(c,x) \to (c, cx, \overline{c}x)$ is realised by an interaction of two gliders [11].

gate that maps $(c, x) \to (c, cx, \overline{c}x)$. It is known that a universal reversible logic gate called a Fredkin gate is composed of two switch gates and two inverse switch gates [3].

In [11] it is shown that three gliders are generated by a head-on collision of two gliders. Using this phenomenon, a glider gun is constructed. Symmetrically, it is also possible to generate two gliders by a collision of three gliders, i.e., one glider is annihilated reversibly.

3.3 Simulating Reversible Logic Element with One-bit Memory in ETPCA 0347

If we want to prove only computational universality of an RCA, it suffices to show that some universal reversible logic gate, such as a Fredkin gate, is embeddable in it. But, if we want to give a concrete configuration that performs some meaningful computation in the RCA, logic gates are not a good choice for it. The reason is as follows.

Consider a logic gate with two or more input lines. Generally, input signals come from places apart from the gate position. Thus, we need some synchronization mechanism so that these signals arrive at the gate at the same time. Since "waiting" for other signals is not possible in a reversible CA, a delay line for the signal should be added to adjust the arrival timing exactly. It needs a cumbersome calculation of signal delays. Moreover, when we construct a sequential machine using a large number of logic gate, it must contain very long feedback loops to keep its state. Hence, its period will be very large, and the whole circuit becomes huge to adjust the timing.

Here, instead of a logic gate, we use a reversible logic element with one-bit memory (RLEM). In an RLEM, an incoming signal interacts only with the internal state of the RLEM rather than other signals. Therefore, the signal can arrive at the RLEM at any time. Hence there are very few problems on the signal timing.

3.3.1 Reversible logic element with memory (RLEM)

A *reversible logic element with memory* (RLEM) is a kind of reversible finite automaton that has an output port as well as an input port. It is a kind of a reversible sequential machine of Mealy type. Formally, a *sequential machine* M is defined as a system $M = (Q, \Sigma, \Gamma, \delta)$, where Q is a finite set of states, Σ and Γ are finite sets of input and output symbols, and $\delta : Q \times \Sigma \to Q \times \Gamma$ is a move function. If δ is injective, it is called a *reversible sequential machine*. An RLEM is defined as a reversible sequential machine that satisfies $|\Sigma| = |\Gamma|$. An RLEM with one-bit memory (i.e., $|Q| = 2$) is particularly important, since it is simple yet powerful. In fact, almost all 2-state RLEMs except only four are universal [10, 13].

We use a specific 2-state 4-symbol RLEM 4-31 where $Q = \{0, 1\}$, $\Sigma = \{a, b, c, d\}$, $\Gamma = \{w, x, y, z\}$, and δ is defined as follows:

$$\delta\ (0, a) = (0, w),\ \delta\ (0, b) = (0, x),\ \delta\ (0, c) = (0, y),\ \delta\ (0, d) = (1, w),$$

$$\delta\ (1, a) = (1, x),\ \delta\ (1, b) = (0, z),\ \delta\ (1, c) = (1, z),\ \delta\ (1, d) = (1, y).$$

In RLEM 4-31, "4" stands for "4-symbol," and "31" is its serial number in the class of 2-state 4-symbol RLEMs. It is universal in the sense any reversible sequential machine can be composed only of it.

The move function δ of RLEM 4-31 is represented in a pictorial form as shown in Fig. 18(a). Two rectangles in the figure correspond to the two states 0 and 1. For each input symbol (output symbol, respectively) of RLEM 4-31, there is a unique input (output) port, to (from) which

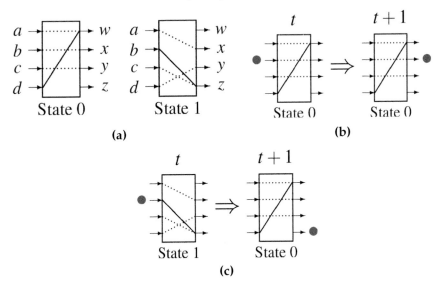

(a)

(b)

(c)

Fig. 18. RLEM 4-31, and its operation examples. (a) Two states of RLEM 4-31. (b) The case that the state does not change, and (c) the case that the state changes.

a signal is given (goes out). In other words, we interpret an RLEM as a reversible sequential machine with "decoded" input/output ports. Hence a signal (or token) must be given at most one input port, and it will appear at most one output port. In the figure, solid and dotted lines show paths of a signal, which gives the input-output relation in each state. If an input signal goes through a dotted line, then the state does not change (Fig. 18(b)). On the other hand, if a signal goes through a solid line, then the state changes (Fig. 18(c)).

The reason why we chose RLEM 4-31 from many RLEMs is as follows. First, as we shall see in Sec. 3.4, reversible Turing machines are compactly realised by RLEM 4-31. Second, RLEM 4-31 is easily implemented in ETPCA 0347, since each state of RLEM 4-31 has only one case where the state changes to the other (i.e., each state has only one solid line as shown in Fig. 18(a)).

3.3.2 Realisation of RLEM 4-31 in ETPCA 0347

RLEM 4-31 is realised in ETPCA 0347 by utilising the fundamental phenomena shown in Sec. 3.2. Figure 19 is the pattern that simulates RLEM 4-31 [8, 9]. Here, a glider is used as a signal. The bounding box of the pattern has no function, but its slits indicate the four input ports and four output ports for a glider. Inside the bounding box there are many blocks and a fin. Blocks are for composing backward-turn, right-turn, and U-turn modules that control the moving direction and the timing of a glider as explained in Sec. 3.2.2. On the other hand, a fin is used to keep the state of the RLEM. Two circles in the middle of the pattern show its possible positions. If the fin is at the lower (upper, respectively) position, then we regard the RLEM is in the state 0 (1). We use the pulling and pushing operations to a fin given in Fig. 16 to change the state. In Fig. 19, the trajectory of the glider in the case of $\delta(1, b) = (0, z)$ is shown. The glider given to the input port b first travels to the lower part of the pattern, and pulls the fin to the southwest direction. Then, the glider moves to the upper part of the pattern, and pushes the fin to the southeast direction. By this, the state of the RLEM changes from 1 to 0. Finally, the glider goes out from the output port z. Though we do not explain other cases than $\delta(1, b) = (0, z)$, they can be verified by the emulator [9] implemented on Golly.

We can see that the pattern in Fig. 19 "directly" simulates RLEM 4-31 in the sense no logic gate is used in it. By this, the size of the pattern that simulates RLEM 4-31 is greatly reduced.

It should be noted that the pattern of Fig. 19 has a very short period 6, which is the period of a fin. When we give an input glider to it, its phase must be 0 when the time is a multiple of 6. Then, the phase of an output glider is adjusted so that the phase becomes 0 when the time is a multiple of 6. By this, we can easily connect many RLEMs to construct a complex

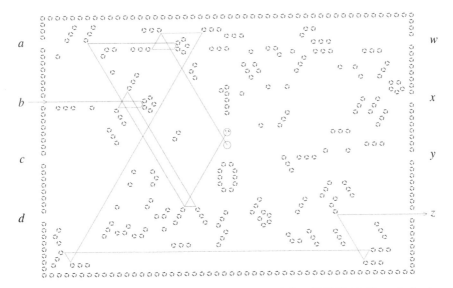

Fig. 19. A pattern that simulates RLEM 4-31 in ETPCA 0347 [8, 9]. Two circles in the middle of the pattern show possible positions of a fin. This pattern shows the state of the RLEM 4-31 is 1, since the fin is at the upper position. Here, a glider is given at the input port b. The path from b to the output port z indicates the trajectory of the glider. By this, the state changes from 1 to 0.

circuit, since the only thing we must do is controlling the trajectory and the phase of a glider.

3.4 Composing Reversible Turing Machines in ETPCA 0347

We use reversible Turing machines as a model of reversible computer, since a general (irreversible) Turing machine is simulated by a reversible one without producing garbage information [1]. Hence they are computationally universal. In this section, we explain that any reversible Turing machine can be realised in the cellular space of ETPCA 0347 rather simply.

3.4.1 Reversible Turing machine (RTM) and its implementation by RLEM 4-31

A *reversible Turing machine* (RTM) is one such that every computational configuration of it has at most one predecessor. Here, we do not give its exact definition. See, e.g. [10, 14] for the definition of an RTM in the quintuple formulation.

It has been shown that any 1-tape 2-symbol RTM can be constructed only of RLEM 4-31 concisely [14]. Figure 20 is an example of a circuit composed of it that simulates an RTM T_{parity}. It is a small RTM having the set of quintuples

$$\{[q_0, 0, 1, R, q_1], [q_1, 0, 1, L, q_{acc}], [q_1, 1, 0, R, q_2], [q_2, 0, 1, L, q_{rej}], [q_2, 1, 0, R, q_1]\}$$

For example, the quintuple $[q_0, 0, 1, R, q_1]$ means that if T_{parity} reads the symbol 0 in the state q_0, then rewrite the symbol to 1, shift the head to the right, and go to the state q_1. Assume a symbol string $0\,1^n\,0$ ($n = 0, 1, \ldots$) is given as an input. Then, T_{parity} halts in the accepting state q_{acc} if and only if n is even, and all the read symbols are complemented. The computing process for the input string 0110 is as follows: $q_0 0110 \Rightarrow 1q_1 110 \Rightarrow 10q_2 10 \Rightarrow 100q_1 0 \Rightarrow 10q_{acc} 01$.

The circuit that simulates T_{parity} given in Fig. 20 consists of two components: a finite control unit (left), and a tape unit (right). The finite control unit is further composed of several state modules, each of which corresponds to a state of T_{parity}. The tape unit is composed of an infinite copies of a memory cell module, which is a vertical array of nine RLEMs. Each memory cell simulates one square of the tape. The top RLEM of a memory cell keeps a tape symbol. The remaining eight RLEMs execute read/write and head-shift commands sent from the finite control. They also keep the head position. Their details are found in [14]. If a particle (or signal) is given to the "Begin" port of Fig. 20, it starts to compute an answer. Its answer will be obtained at "Accept" or "Reject" port.

3.4.2 Simulating RTMs in ETPCA 0347

Since RLEM 4-31 is implemented as a pattern given in Fig. 19, it is easy to obtain a configuration that simulates T_{parity}. Putting copies of the pattern of Fig. 19 at the positions corresponding to the RLEMs in Fig. 20, and connecting them appropriately, we have a complete configuration of ETPCA 0347 that simulates T_{parity}. Figure 21 shows the configuration

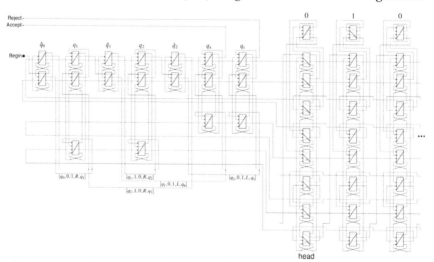

Fig. 20. A circuit composed of RLEM 4-31 that simulates the RTM T_{parity} [14].

Fig. 21. A configuration of ETPCA 0347 in Golly that simulates
the circuit of Fig. 20 [9].

simulated in Golly. In this figure, each small parallelogram shows the
pattern of RLEM 4-31 (note that the pattern in ETPCA 0347 that simulates
RLEM 4-3 is slanted in Golly, since a pair of equilateral triangles are
simulated by one square cell of Golly). Giving a glider to "Begin" port,
its computation starts. Computing process of T_{parity} in ETPCA 0347 can
be seen by the emulator [9] on Golly. A slightly more complex example
of an RTM is also given in [9]. Actually, for any RTM we can compose
a configuration of ETPCA 0347 that simulates the RTM in a systematic
manner.

4. Concluding Remarks

In this chapter, we investigated the problem of how we can compose a
computing machine from a collection of a large number of cells that obey
a quite simple reversible law. We showed it is indeed possible to find a
pathway from a reversible law to reversible computers using ETPCA
0347 shown in Fig. 2. There are following four stages in the pathway.
First, useful basic patterns (or objects) in the cellular space of ETPCA
0347 are identified. Second, various fundamental phenomena that appear
in the interactions of basic patterns are found. Third, a suitable reversible
logic element RLEM 4-31 is supposed to compose reversible computers,
and it is realised by utilising fundamental phenomena in ETPCA 0347.
Finally, a reversible Turing machines is composed from RLEM 4-31, and
it is embedded in the cellular space of ETPCA 0347. Reversible computers
are thus constructed in a hierarchical and systematic manner along this
pathway.

Besides ETPCA 0347, it has already been shown that the reversible
ETPCAs 0137, 0157 and 0267 are computationally universal, since a
Fredkin gate is realised in each of their cellular space [6, 7, 10]. We think
it is also possible to compactly realise reversible Turing machines in these
ETPCAs as in the case of ETPCA 0347. Thus it is left for future study.

References

1. C.-H. Bennett. Logical reversibility of computation. IBM J. Res. Dev. 17, 525–532 (1973).
2. E. Berlekamp, J. Conway and R. Guy. Winning Ways for Your Mathematical Plays, Vol. 2. Academic Press, New York (1982).
3. E. Fredkin and T. Toffoli. Conservative logic. Int. J. Theoret. Phys. 21, 219–253 (1982).
4. M. Gardner. Mathematical games: the fantastic combinations of John Conway's new solitaire game "life". Sci. Am. 223(4), 120–123 (1970).
5. M. Gardner. Mathematical games: on cellular automata, self-reproduction, the Garden of Eden and the game "life". Sci. Am. 224(2), 112–117 (1971).
6. K. Imai and K. Morita. A computation-universal two-dimensional 8-state triangular reversible cellular automaton. Theoret. Comput. Sci. 231, 181–191 (2000).
7. K. Morita. Universality of 8-state reversible and conservative triangular partitioned cellular automata. *In*: S. El Yacoubi et al. (eds.), ACRI 2016, LNCS 9863, pp. 45–54 (2016).
8. K. Morita. Making reversible Turing machines in a reversible elementary triangular partitioned cellular automaton. Proc. AUTOMATA 2017 Exploratory Papers, pp. 77–84 (2017).
9. K. Morita. Reversible world: data set for simulating a reversible elementary triangular partitioned cellular automaton on Golly. Hiroshima University Institutional Repository, http://ir.lib.hiroshima-u.ac.jp/00042655 (2017).
10. K. Morita. Theory of Reversible Computing. Springer, Tokyo (2017).
11. K. Morita. A universal non-conservative reversible elementary triangular partitioned cellular automaton that shows complex behavior. Natural Computing, 18, 413-428 (2019). Slides with movies of computer simulation: Hiroshima University Institutional Repository, http://ir.lib.hiroshima-u.ac.jp/00039321.
12. K. Morita and M. Harao. Computation universality of one-dimensional reversible (injective) cellular automata. Trans. IEICE Japan, E72, 758–762 (1989).
13. K. Morita, T. Ogiro, A. Alhazov and T. Tanizawa. Non-degenerate 2-state reversible logic elements with three or more symbols are all universal. J. Multiple-Valued Logic and Soft Computing 18, 37–54 (2012).
14. K. Morita and R. Suyama. Compact realization of reversible Turing machines by 2-state reversible logic elements. *In*: O.H. Ibarra, L. Kari, S. Kopecki (eds.), Proc. UCNC 2014, LNCS 8553, pp. 280–292 (2014). Slides with figures of computer simulation: Hiroshima University Institutional Repository, http://ir.lib.hiroshima-u.ac.jp/00036076.
15. A. Trevorrow and T. Rokicki. Golly: an open source, cross-platform application for exploring Conway's Game of Life and other cellular automata. http://golly.sourceforge.net/ (2005).
16. S. Wolfram. Theory and Applications of Cellular Automata. World Scientific Publishing (1986).
17. S. Wolfram. A New Kind of Science. Wolfram Media Inc. (2002).

7

Rough Sets over Social Networks

Krzysztof Pancerz* and Piotr Grochowalski

College of Natural Sciences, University of Rzeszow, Poland

1. Introduction

In general, the world of social behaviour cannot be described unambiguously and precisely. This is caused due to an entanglement of social interactions between categories distinguished in large social environments. In this chapter, we will show one of the possibilities of how to assess (measure) ambiguities or imprecisions that have to be dealt with if we wish to determine some categories on the basis of connections (social interactions) with other categories in social networks. In large social structures with plenty of interactions, concepts cannot be defined unambiguously or precisely. In this case, they can be described by certain approximations induced by means of connections. Therefore, we propose to use rough sets to deal with the problem of ambiguity or imprecision. Rough sets are an appropriate tool to deal with rough (ambiguous, imprecise) categories. Theoretical foundations given in the chapter are supplemented with a real-life examples concerning the assessment and cohesion of co-authorships between research teams.

To outline the presented approach, we can generally describe the problem dealt with as follows:

Input: A social network SN presenting some social interactions and distinguished social categories (groups, teams, etc.) built over the set of actors in SN.

Goal: To determine the cohesion of social interactions between social categories identified in the social network SN.

Methodology: Rough sets defined over the social network SN expressing some ambiguities or imprecisions induced by social interactions.

---

*Corresponding author: kpancerz@ur.edu.pl

In the remaining part of the chapter, we present:

- related works concerning, among others, applications of rough sets for graph (network) structures (Sec. 2),
- basic notions concerning collaboration graphs (Sec. 3.1) and rough sets (Sec. 3.2),
- the theoretical description of the proposed methodology (Sec. 4),
- the practical example of real-life applications of the proposed methodology (Sec. 5),
- conclusions and the possible directions of further research (Sec. 6).

2. Related Works

Plenty of beings, both animate and inanimate, in the world are connected with each other. Networks (structures with connected beings) are everywhere, from the Internet, to social networks, and genetic networks [1]. In this chapter, we will be interested in social networks playing an increasingly important role in perceiving and creating the world around us. In recent years, social networks have become a powerful, and more formal tool in describing the social behaviour used in many research areas, from the slightly more generalised ideas of, sociology, biology, zoology, etc. to the more specific ideas of, bacteriology, epidemiology, bibliometrics, communication theory, etc.

In general, a social network [10], as a more formal tool, is a social structure made up of:

- a set of actors,
- a set of dyadic connections between the actors.

Actors can be either individuals or organizations. Dyadic connections represent social interactions. A special kind of a social structure is the scientific community, treated generally or considered in a specific scientific area of interest. In this case, the basic social interaction is a scientific collaboration implemented, among others, in the form of papers published jointly. Such a collaboration can be described by the so-called author collaboration graph (cf. [2]). We recall basic definitions in Sec. 3.1.

A key ability is to be able to determine certain properties of social networks or to discover patterns hidden in these networks. A large toolkit (including approaches based mainly on statistics and graph theory) is available for characterizing the structures and dynamics of complex networks (cf. [3]). Social networks together with complex networks are considered in the nascent field of network science [1].

In this chapter, we are interested in the assessment of ambiguities or imprecisions that we have to be dealt with if we determined categories on the basis of connections (social interactions) with other categories in social networks. It is natural that, in large social structures with plenty of interactions, concepts cannot be defined unambiguously, precisely, etc. This fact makes it necessary to use a tool enabling us to deal with the problem of ambiguity or imprecision. One of the possibilities is to use rough sets as proposed by Z. Pawlak [7] as a tool to deal with rough (ambiguous, imprecise) categories.

In our earlier research, we have used rough sets defined over different graph (network) structures. The most significant ones were:

- complex networks – rough sets were used to analyse certain properties of complex networks modeling eye-tracking sequences [4],
- transition systems – rough sets were used to assess ambiguities in plas- modium propagation [6].

Several approaches to rough set based descriptions of plasmodium propagation have been described in [8]. In our current research which is described in this chapter, we adopt and extend the approaches used for complex networks and transition systems to social networks. Now, in general, rough sets are defined over social networks.

3. Basic Notions

The formal description of the proposed methodology for determining the cohesion of social interactions between social categories (presented in Sec. 4) uses basic notions recalled in the following subsections.

3.1 Collaboration Graphs

In this chapter, we show how to define rough sets over social networks. In the presented case, an underlying structure is an author collaboration graph. The author collaboration graph (cf. [2]) is an undirected graph $ACG = (A, C)$, where A is the set of nodes corresponding to authors of papers and C is the set of edges corresponding to co-authorships of joint papers. An edge connects two authors if they have published at least one joint paper, with or without other co-authors.

We can also consider the weighted author collaboration graph that is a weighted undirected graph $WACG = (A, C, v)$, where A and C are defined previously, as well as $v : C \rightarrow \{1, 2, \dots \}$ is the weight function assigning a positive integer (determining how many papers the two authors have published with or without other co-authors) to each edge.

3.2 Rough Sets

Rough sets were defined by Z. Pawlak [7] as a tool to deal with rough (ambiguous, imprecise) categories. The main idea is to approximate a given set by a pair of sets, called the lower and upper approximation of this set. Some sets cannot be exactly defined. If a given set X is not exactly defined, then we employ two exact sets (the lower and the upper approximation of X) that define X roughly (approximately). Originally, rough sets were defined on the basis of indiscernibility relation of objects in some universe of discourse. Next, different generalizations have been proposed. In general, rough sets can be defined on the basis of any binary relation between objects (see [11]).

In this section, we recall basic notions used in rough set theory that are necessary to understand the description of methodology shown in Sec. 4.

Let $U \neq \emptyset$ be a finite set of objects we are interested in. U is called the universe. Let R be any binary relation over U. With each subset $X \subseteq U$ and any binary relation R over U, we associate two subsets:

- $\underline{R}(X) = \{u \in U : R(u) \subseteq X\}$,
- $\overline{R}(X) = \{u \in U : R(u) \cap X \neq \emptyset\}$,

called the R-lower and R-upper approximation of X, respectively. A set

$$BN_R(X) = \overline{R}(X) - \underline{R}(X) \tag{1}$$

is called the R-boundary region of X. If $BN_R(X) = \emptyset$, then X is sharp (exact) with respect to R. Otherwise, X is rough (inexact).

Roughness of a set X can be characterized numerically. To this end, the accuracy of approximation of X with respect to R is defined as:

$$\alpha_R(X) = \frac{card(\underline{R}(X))}{card(\overline{R}(X))} \tag{2}$$

where *card* denotes the cardinality of the set and $X \neq \emptyset$.

The definitions given earlier are based on the standard definition of set inclusion. Let U be the universe and $A, B \subseteq U$. The standard set inclusion is defined as

$$A \subseteq B \text{ if and only if } \underset{u \in A}{\forall} u \in B. \tag{3}$$

Sometimes, this definition becomes too restrictive and rigorous. W. Ziarko proposed in [12] some relaxation of the original rough set approach which is called the Variable Precision Rough Set Model (VPRSM). The VPRSM approach is based on the notion of relaxed set inclusion. Let U be the universe, $A, B \subseteq U$, and $0 \leq \beta < 0.5$. The relaxed set inclusion is defined as

$$A \overset{\beta}{\subseteq} B \text{ if and only if } 1 - \frac{card(A \cap B)}{card(A)} \leq \beta \qquad (4)$$

where *card* denotes the cardinality of the set. $A \overset{\beta}{\subseteq} B$ means that a specified number of elements belonging to A belongs also to B. One can see that if $\beta = 0$, then the relaxed set inclusion becomes the standard set inclusion.

In our approach, we will use slightly modified definitions of approximations based on the relaxed set inclusion, i.e.:

- $\underline{R}^{\beta}(X) = \left\{ u \in U : R(u) \overset{\beta}{\subseteq} X \right\}$,

- $\overline{R}^{\beta}(X) = \left\{ u \in U : 1 - \frac{card(R(u) \cap X)}{card(R(u))} \leq 1 - \beta \right\}$,

called the R_{β}-lower and R_{β}-upper approximation of X, respectively, where $0 \leq \beta \leq 0.5$.

The approach presented in this chapter refers to a general framework for the study of approximation using the notion of neighborhood systems proposed by T.Y. Lin (cf. [5]).

4. Cohesion of Co-authorships Between Research Teams

The approach presented in this section is an adaptation and extension of our previous approaches presented in [4] (where rough set theory was used to analyze some of the properties of complex networks modeling eye-tracking sequences) and [6] (where rough set theory was used to assess ambiguities in plasmodium propagation). Now, generally speaking, we define rough sets over social networks. As an example, we use rough set theory to assess the cohesion of co-authorships between research teams.

The underlying data structure in the proposed approach is the author collaboration graph $ACG = (A, C)$ or its extension – the weighted author collaboration graph $WACG = (A, C, v)$ (see Sec. 3.1). In fact, the rough sets are defined over these structures.

The nodes corresponding to authors of papers can be grouped, for example, a group can represent a research team in a given scientific institution.

Let $T = \{\tau_1, \tau_2, \ldots, \tau_t\}$ be a set of all identified research teams. We use the following notation:

- $A = \{A_{\tau 1}, A_{\tau 2}, \ldots, A_{\tau t}\}$ denotes the family of sets of nodes corresponding to those teams.
- A^Θ denotes a set of nodes corresponding to authors not belonging to any identified research team.

Let $A = A_{\tau 1} \cup A_{\tau 2} \cup \cdots \cup A_{\tau t}$ be the set of authors belonging to research teams.

For each node $a \in A_{\tau 1} \cup A_{\tau 2} \cup \cdots \cup A_{\tau t}$, we identify its inter-team neighbourhood as:

$$ITN(a) = \{a' : (a, a') \in C \wedge \underset{\tau \in T}{\exists} (a' \in A_{\tau} \wedge a \notin A_{\tau})\} \tag{5}$$

To define a measure, derived from rough set theory, for assessing the cohesion of co-authorships between research teams, we use the lower and upper inter-team neighbourhood approximations (cf. [4] and [6]).

Let $ACG = (A, C)$ be an author collaboration graph and $\tau_i, \tau_j \in T$ be two distinguished research teams. The lower inter-team neighbourhood approximation $\underline{ITN}(\tau_j | \tau_i)$ of team τ_j by team τ_i is defined as:

$$\underline{ITN}(\tau_j | \tau_i) = \{a \in A_{\tau_i} : ITN(a) \neq \emptyset \wedge ITN(a) \subseteq A_{\tau_j}\} \tag{6}$$

The upper inter-team neighbourhood approximation $\overline{ITN}(\tau_j | \tau_i)$ of team τ_j by team τ_i is defined as:

$$\overline{ITN}(\tau_j | \tau_i) = \{a \in A_{\tau_i} : ITN(a) \cap A_{\tau_j} \neq \emptyset\} \tag{7}$$

One can identify the following interpretation of approximations, lower and upper. The lower inter-team neighbourhood approximation $\underline{ITN}(\tau_j | \tau_i)$ of team τ_j by team τ_i determines the set of all authors from team τ_i having the co-authorships of papers with authors from team τ_j only. The upper inter-team neighbourhood approximation $\overline{ITN}(\tau_j | \tau_i)$ of team τ_j by team τ_i determines the set of authors from team τ_i having some co-authorships of papers with authors from team τ_j.

The cohesion measure of co-authorships between two research teams can be defined on the basis of lower and upper inter-team neighbourhood approximations, like the accuracy of approximation is defined in rough set theory, i.e.:

$$\chi_{ITN}(\tau_j | \tau_i) = \frac{card(\underline{ITN}(\tau_j | \tau_i))}{card(\overline{ITN}(\tau_j | \tau_i))} \tag{8}$$

Further, this cohesion measure will be called the basic cohesion measure of co-authorships between two research teams. One can see that if $\chi_{ITN}(\tau_j | \tau_i) = 1$, then the co-authorships between two research teams are the most coherent ones.

It is worth noting that the cohesion measure of co-authorships between two research teams is not symmetrical, i.e., in general:

$$\chi_{ITN}(\tau_j | \tau_i) \neq \chi_{ITN}(\tau_j | \tau_i), \tag{9}$$

for $i, j = 1, 2, \ldots, t$ and $i \neq j$.

The cohesion measure defined earlier can be modified in many different ways. We are interested in two modification ways.

Firstly, we can use modified definitions of the lower and upper inter-team neighbourhood approximations coming from the Variable Precision Rough Set Model (VPRSM) (see Sec. 3.2).

Let $ACG = (A, C)$ be an author collaboration graph, $\tau_i, \tau_j \in T$ be two distinguished research teams, and $0 \le \beta \le 0.5$. The β-lower inter-team neighbourhood approximation $\underline{ITN}^{\beta}(\tau_j \,|\, \tau_i)$ of team τ_j by team τ_i is defined as:

$$\underline{ITN}^{\beta}(\tau_j \,|\, \tau_i) = \left\{ a \in A_{\tau_i} : ITN(a) \ne \emptyset \wedge ITN(a) \overset{\beta}{\subseteq} A_{\tau_j} \right\} \tag{10}$$

The β-upper inter-team neighbourhood approximation $\overline{ITN}^{\beta}(\tau_j \,|\, \tau_i)$ of team τ_j by team τ_i is defined as:

$$\overline{ITN}^{\beta}(\tau_j \,|\, \tau_i) = \left\{ a \in A_{\tau_i} : 1 - \frac{card(ITN(a) \cap A_{\tau_j})}{card(ITN(a))} \le 1 - \beta \right\}. \tag{11}$$

Now, one can identify the following interpretation of the lower approximation. The lower inter-team neighbourhood approximation $\underline{ITN}(\tau_j \,|\, \tau_i)$ of team τ_j by team τ_i determines the set of all authors from team τ_i having at least the specified number of papers written jointly with authors from team τ_j.

The modified cohesion measure of co-authorships between two research teams has the form:

$$\chi_{ITN}^{\beta}(\tau_j \,|\, \tau_i) = \frac{card(\underline{ITN}^{\beta}(\tau_j \,|\, \tau_i))}{card(\overline{ITN}^{\beta}(\tau_j \,|\, \tau_i))} \tag{12}$$

for $0 \le \beta \le 0.5$. It is worth noting that the modified cohesion measure is the relaxed basic cohesion measure $\chi_{ITN}(\tau_j \,|\, \tau_i)$. Therefore, the measure $\chi_{ITN}^{\beta}(\tau_j \,|\, \tau_i)$ will be called the relaxed cohesion measure of co-authorships between two research teams.

Secondly, we can use a weighted author collaboration graph $W\,ACG = (A, C, v)$ (see Sec. 3.1) instead of an author collaboration graph $ACG = (A, C)$. The weights of edges modify the basic cohesion measure $\chi_{ITN}(\tau_j \,|\, \tau_i)$.

The weighted cohesion measure $\overset{\bullet}{\chi}_{ITN}(\tau_j \,|\, \tau_i)$ of co-authorships between two research teams is defined as:

$$\overset{\bullet}{\chi}_{ITN}(\tau_j \,|\, \tau_i) = \frac{card(\underline{ITN}(\tau_j \,|\, \tau_i))}{card(\underline{ITN}(\tau_j \,|\, \tau_i) + f\,card(BN_{ITN}(\tau_j \,|\, \tau_i))}, \tag{13}$$

where

$$BN_{ITN}(\tau_j \,|\, \tau_i) = \overline{ITN}(\tau_j \,|\, \tau_i) - \underline{ITN}(\tau_j \,|\, \tau_i)) \tag{14}$$

and

$$f\,card(BN_{ITN}(\tau_j \mid \tau_i)) = \sum_{a \in BN_{ITN}(\tau_j \mid \tau_i)} \left(1 - \frac{\sum\limits_{a' \in A_{\tau_j}} v((a, a'))}{\sum\limits_{a' \in A - A_{\tau_j}} v((a, a'))} \right). \qquad (15)$$

One can see that $f\,card$ is a some kind of a fuzzy cardinality of a set. In case of the basic cohesion measure, all nodes from the upper inter-team neighbourhood approximation are taken entirely to determine the cardinality of this upper approximation. Now, the node a from the boundary region is not taken as a whole element. It is taken partially according to the ratio of weights of edges connecting a with nodes corresponding to authors from the approximated research team and weights of edges connecting a with nodes corresponding to authors from all research teams excluding the research team of the author represented by a. This ratio determines how relatively strong are the edges causing that a is in the boundary region.

It is worth noting that the combination of the relaxed cohesion measure $\chi_{ITN}^{\beta}(\tau_j \mid \tau_i)$ and the weighted cohesion measure $\overset{\bullet}{\chi}_{ITN}(\tau_j \mid \tau_i)$ is also possible.

5. Real-life Application

The most recognizable example of social interactions within the scientific community is conducting common research. The measurable effect of research is a set of common scientific publications. On the Internet, there are a lot of bibliographical databases available, both general and specialized. We can derive from examining such databases information about co-authorships between individual authors or whole research teams.

In our research, we use the Rough Set Database System (RSDS) [9] developed from 2002 by the Research Group on *Rough Sets and Petri Nets*, functioning at the University of Rzeszʹow in cooperation with the International Rough Set Society. The RSDS database is designed mainly for including bibliographic descriptions of publications on rough set theory, its applications, and related fields. This database is also an experimental environment for research related to processing of bibliographic data, especially with the use of the domain knowledge and for research related to information retrieval. Currently, information about over 37000 publications written by over 42000 authors is included in RSDS. The RSDS system is available at http://rsds.ur.edu.pl (August 2019).

Let us consider, manual calculations explaining the proposed approach, a small piece of the weighted author collaboration graph generated from the RSDS database. This graph is presented in Table 1 in the form of the weighted adjacency matrix. For simplicity, this piece will be denoted as $W\,ACG = (A, C, v)$. For this graph:

- the set of nodes representing authors
    A = {*Grochowalski, Knap, Lew, Paja, Pancerz, Rzasa, Suraj*},
- the set of edges representing co-authorships of joint papers
    C = {(*Grochowalski, Lew*), (*Grochowalski, Pancerz*), (*Grochowalski, Suraj*), (*Knap, Paja*), (*Lew, Grochowalski*), (*Lew, Suraj*), (*Paja, Knap*), (*Paja, Pancerz*), (*Pancerz, Grochowalski*), (*Pancerz, Paja*), (*Pancerz, Suraj*), (*Pancerz, Szkola*), (*Rzasa, Suraj*), (*Suraj, Grochowalski*), (*Suraj, Lew*), (*Suraj, Pancerz*), (*Suraj, Rzasa*), (*Szkola, Pancerz*)}.
- the weight function determining how many papers the two authors have published with or without other co-authors has values as presented in Table 1.

Table 1. The weighted adjacency matrix representing the weighted author collaboration graph $W\,ACG = (A, C, v)$

Authors	Grochowalski	Knap	Lew	Paja	Pancerz	Rzasa	Suraj	Szkola
Grochowalski	0	0	2	0	6	0	24	0
Knap	0	0	0	2	0	0	0	0
Lew	2	0	0	0	0	0	2	0
Paja	0	2	0	0	1	0	0	0
Pancerz	6	0	0	1	0	0	38	1
Rzasa	0	0	0	0	0	0	13	0
Suraj	24	0	2	0	38	13	0	0
Szkola	0	0	0	0	1	0	0	0

We will take into consideration the following research teams over the set A of authors:

- $Team_1$ = {*Grochowalski, Lew, Suraj*},
- $Team_2$ = {*Pancerz, Szkola*},
- $Team_3$ = {*Knap, Paja*},
- $Team_4$ = {*Rzasa*}.

Our goal in this example is to determine the cohesion measure of co-authorships between research teams $Team_1$ and $Team_2$. More exactly, we will find inter-team neighbourhood approximations of $Team_2$ by $Team_1$.

For calculations of the inter-team neighbourhood approximations, we omit in the weighted adjacency matrix those entries which describe the numbers of papers published by two authors belonging to $Team_1$. The reduced weighted adjacency matrix is shown in Table 2.

Table 2. The reduced (with respect to $Team_1$) weighted adjacency matrix

Authors	Grochowalski	Knap	Lew	Paja	Pancerz	Rzasa	Suraj	Szkola
Grochowalski	0	0	0	0	6	0	0	0
Knap	0	0	0	2	0	0	0	0
Lew	0	0	0	0	0	0	0	0
Paja	0	2	0	0	1	0	0	0
Pancerz	6	0	0	1	0	0	38	1
Rzasa	0	0	0	0	0	0	13	0
Suraj	0	0	0	0	38	13	0	0
Szkola	0	0	0	0	1	0	0	0

It is easy to see that the inter-team neighbourhoods, for the authors from team $Team_1$, are as follows:

- ITN (Grochowalski) = {Pancerz},
- ITN (Lew) = $\varnothing$,
- ITN (Suraj) = {Pancerz, Rzasa}.

Now we can determine the lower and upper inter-team neighbourhood approximations $\underline{ITN}$ $(Team_2 \mid Team_1)$ and $\overline{ITN}$ $(Team_2 \mid Team_1)$, respectively:

- $\underline{ITN}$ $(Team_2 \mid Team_1)$ = {Grochowalski} because
  - ITN (Grochowalski) $\subseteq Team_2$,
  - ITN (Lew) = $\varnothing$,
  - ITN (Suraj) $\not\subseteq Team_2$.
- $\overline{ITN}$ $(Team_2 \mid Team_1)$ = {Grochowalski, Suraj} because
  - ITN (Grochowalski) $\cap Team_2$ = {Pancerz} $\neq \varnothing$,
  - ITN (Lew) $\cap Team_2$ = $\varnothing$,
  - ITN (Suraj) $\cap Team_2$ = {Pancerz} $\neq \varnothing$.

Hence, the basic cohesion measure of co-authorships between research teams $Team_1$ and $Team_2$:

$$\chi_{ITN}(Team_2 \mid Team_1) = \frac{card(\underline{ITN}(Team_2 \mid Team_1))}{card\overline{ITN}(Team_2 \mid Team_1)} = 0.5$$

If we use the relaxed set inclusion, we obtain for $\beta = 0.5$:

- $\underline{ITN}^{0.5}$ $(Team_2 \mid Team_1)$ = {Grochowalski, Suraj} because
  - ITN (Grochowalski) $\overset{0.5}{\subseteq} Team_2$,
  - ITN (Lew) = $\varnothing$,
  - ITN (Suraj) $\overset{0.5}{\subseteq} Team_2$.
- $\overline{ITN}^{0.5}$ $(Team_2 \mid Team_1)$ = {Grochowalski, Suraj} because

  $$\frac{card(ITN(Grochowalski) \cap Team_2)}{card(ITN(Grochowalski))} = 1,$$

$$- \quad \frac{card(ITN(Lew) \cap Team_2)}{card(ITN(Lew))} = 0,$$

$$- \quad \frac{card(ITN(Suraj) \cap Team_2)}{card(ITN(Suraj))} = 0.5.$$

Hence, the relaxed cohesion measure of co-authorships between research teams $Team_1$ and $Team_2$:

$$\chi_{ITN}^{0.5}(Team_2 \mid Team_1) = \frac{card(\underline{ITN}^{0.5}(Team_2 \mid Team_1))}{card(\overline{ITN}^{0.5}(Team_2 \mid Team_1))} = 1.0.$$

Finally, we can calculate the weighted cohesion measure of co-authorships between research teams $Team_1$ and $Team_2$. In this case, we use the numbers of co-authorship publications as they are shown in Table 1. It is worth noting that, for the basic cohesion measure and the relaxed cohesion measure, the adjacency matrix was considered as the binary one (0 for the weight function equal to 0 and 1 for the weight function greater than 0).

The boundary region $BN_{ITN}(Team_2 \mid Team_1) = \{Suraj\}$. The fuzzy cardinality for $BN_{ITN}(Team_2 \mid Team_1)$ is as follows:

$$f\,card\,(BN_{ITN}((Team_2 \mid Team_1))) = 1 - \frac{v((Suraj, Pancerz))}{v((Suraj, Pancerz)) + v((Suraj, Rzasa))}$$

$$= 0.2549.$$

Hence, the weighted cohesion measure of co-authorships between research teams $Team_1$ and $Team_2$:

$$\chi_{ITN}^{\bullet}(Team_2 \mid Team_1) =$$

$$\frac{card(\underline{ITN}(Team_2 \mid Team_1))}{card(\underline{ITN}(Team_2 \mid Team_1)) + f\,card(BN_{ITN}(Team_2 \mid Team_1))} = \frac{1}{1 + 0.2549}$$

$$= 0.7969.$$

The presented methodology has been implemented in the R environment. To calculate the lower and upper inter-team neighbourhood approximations we can use the following script (*collaboration.matrix* is a matrix including the considered adjacency matrix, *indexes* is a vector of all considered indexes, x and y are used to determine the approximating and approximated research teams, respectively):

Calculation of the lower and upper inter-team neighbourhood approximations

```
team_1<-c(1,3,7); team_2<-c(5,8); team_3<-c(2,4); team_4<-c(6);
x<-team_1; y<-team_2;
lower<-c(); upper<-c();

for(i in 1:length(x))
{
  itn<-c();

  for(j in 1:length(indexes))
  {
    if(!(j %in% x))
    { if(collaboration.matrix[x[i], j]>0) {itn<-c(itn, j); } }
  }

  if(length(itn)>0 & all(itn %in% y)) { lower<-c(lower, x[i]); }
  if(length(intersect(itn, y))>0) { upper<-c(upper, x[i]); }
}
```

Next, the basic and weighted cohesion measures of co-authorships between research teams can be calculated:

Calculation of the basic and weighted cohesion measures of co-authorships between research teams

```
fuzzy.card<-0;

for(i in 1:length(upper))
{
  if(!(upper[i] %in% lower))
  {
    target.weights<-0; all.weights<-0;

    for(j in 1:length(indexes))
    {
      if(j %in% y)
      { target.weights<-target.weights+collaboration.atrix[upper[i],j]; }

      if(!(j %in% x))
      { all.weights<-all.weights+collaboration.matrix[upper[i],j]; }
    }
    fuzzy.card<-fuzzy.card+(1-target.weights/all.weights);
  }
}

basic.cohesion<-length(lower)/length(upper);
weighted.cohesion<-length(lower)/(length(lower)+fuzzy.card);
```

---

6. Conclusions

As an example of application of rough sets defined over social networks, the problem of determining the cohesion measure of co-authorships between research teams has been considered. One can see that the presented methodology can be used for any social interactions described by social networks. The im- portant thing is to identify the social categories which would be approximated. Approximations are induced from information about social interactions. Hence, we can describe the cohesion (in some cases describing the strength) of a collaboration between both animate (people, bacteria, or other live creatures) and inanimate (ICT devices, sensors, etc.) beings defined in the considered environments. The important research problem is to be able to deal directly with not only dyadic connections representing social interactions, but also with triadic or, in general, polyadic connections.

References

1. A.-L. Barabási. Network Science. Cambridge University Press (2016).
2. V. Batagelj and A. Mrvar. Some analyses of Erdös collaboration graph. Social Networks 22(2), 173–186 (2000).
3. S. Boccaletti, V. Latora, Y. Moreno, M. Chavez and D.-U. Hwang. Complex networks: structure and dynamics. Physics Reports 424, 175–308 (2006).
4. B. Jaskuła, J. Szkoła, K. Pancerz and A. Derkacz. Eye-tracking data, complex networks and rough sets: an attempt toward combining them. *In*: J. Suzuki, T. Nakano and H. Hess (eds), Proceedings of the 9th International Conference on Bio-inspired Information and Communications Technologies (BICT'2015), pp. 167–173. New York City, New York, USA (2015).
5. T.Y. Lin. Topological and fuzzy rough sets. *In*: R. Słowiński (ed.), Intelligent Decision Support: Handbook of Applications and Advances of the Rough Sets Theory, pp. 287–304. Springer Netherlands, Dordrecht (1992).
6. K. Pancerz. Quantitative assessment of ambiguities in plasmodium propagation in terms of complex networks and rough sets. *In*: T. Nakano and A. Compagnoni (eds), Proceedings of the 10th EAI International Conference on Bio-inspired Information and Communications Technologies (BICT'2017), Hoboken, USA, pp. 63–66 (2017).
7. Z. Pawlak. Rough Sets. Theoretical Aspects of Reasoning about Data. Kluwer Academic Publishers, Dordrecht (1991).
8. A. Schumann and K. Pancerz. High-Level Models of Unconventional Computations: A Case of Plasmodium. Springer International Publishing (2019).
9. Z. Suraj and P. Grochowalski. About new version of RSDS system. Fundamenta Informaticae 135(4), 503–519 (2014).
10. S. Wasserman and K. Faust. Social Network Analysis: Methods and Applications. Cambridge University Press (1994).

11. Y.Y. Yao, S.K.M. Wong and T.Y. Lin. A review of rough set models. Rough Sets and Data Mining: Analysis of Imprecise Data, pp. 47–75. Kluwer Academic Publishers, Dordrecht (1997).
12. W. Ziarko. Variable precision rough set model. Journal of Computer and System Sciences 46(1), 39–59 (1993).

8

Logical Functions as an Idealization of Swarm Behavior

Andrew Schumann

University of Information Technology and Management in Rzeszow, Sucharskiego 2, 35-225 Rzeszow, Poland
Email: andrew.schumann@gmail.com

1. Introduction

The idea to design some computers on particles of swarms by using their stable patterns is quite old. Based on this idea different logical and then arithmetic functions were implemented by using the typical reactions of swarms to outer stimuli. For example, it was proposed to have computers consisting of ant colonies [2] and computers composed of plasmodia of *Physarum polycephalum* (a giant multinucleated amoeba) [1, 5, 6], etc. In this way, logical functions are made implementable within swarm patterns. This raises a question that requires serious consideration—how natural are these logical formalizations of swarm reactions? In other words, do swarms realize some logical functions themselves or are these functions implemented on them artificially? In this chapter, a hypothesis is presented on the assumption that the basic reactions of swarms occur due to realization of some logical functions by their nature. Hence, we assume that logic is absolutely natural for swarms.

Each swarm reacts to outer signals by means of decentralized reactions of its members. There are two kinds of outer signals: *attractants* which attract swarm members and *repellents* which repel swarm members. If we deal with many signals, then reactions to them can be naturally represented as logical functions. In Section 2, two-valued logical functions of the following types are defined: logical functions of propositional logic, logical functions of first-order logic of unary predicates, and logical functions of modal logic. In Section 3, the concept of duality of logical functions is examined, which allows us to order logical functions. In Section 4, logical functions are defined on swarm patterns. And in Section 5, the logical duality is considered natural for swarm basic reactions.

2. Two-valued Logical Functions

Suppose that $\mathfrak{I}$ be a standard set of propositional formulas containing propositional variables $p_1, \ldots, p_k$ and their propositional superpositions by using negation $\neg$, conjunction $\wedge$, and disjunction $\vee$. A *two-valued logical meaning m* is defined as a mapping from the set of propositional formulas $\mathfrak{I}$ to $\{\top, \bot\}$, where $\top$ means 'truth' and $\bot$ means 'falsehood', such that:

- let $p \in \mathfrak{I}$ be a propositional variable, then $m(p)$ can give either a value $\top$ or $\bot$;
- let $\neg\varphi \in \mathfrak{I}$ be a negation of propositional formula φ, then $m(\neg\varphi) = \neg m(\varphi)$, where $\neg m(\varphi) = \bot$ if and only if $m(\varphi) = \top$ and $\neg m(\varphi) = \top$ if and only if $m(\varphi) = \bot$;
- let $(\varphi \wedge \psi) \in \mathfrak{I}$ be a conjunction of two propositional formulas φ and ψ, then $m(\varphi \wedge \psi) = (m(\varphi) \wedge m(\psi))$, where $(m(\varphi) \wedge m(\psi)) = \top$ if and only if $m(\varphi) = \top$ and $m(\psi) = \top$, otherwise $(m(\varphi) \wedge m(\psi)) = \bot$;
- let $(\varphi \vee \psi) \in \mathfrak{I}$ be a conjunction of two propositional formulas φ and ψ, then $m(\varphi \vee \psi) = (m(\varphi) \vee m(\psi))$, where $(m(\varphi) \vee m(\psi)) = \bot$ if and only if $m(\varphi) = \bot$ and $m(\psi) = \bot$, otherwise $(m(\varphi) \vee m(\psi)) = \top$.

An *n*-place *two-valued logical function f* is defined as a mapping from $\mathfrak{I}^n$ to $\mathfrak{I}$ such that $m(f(\varphi_1, \ldots, \varphi_n)) = f(m(\varphi_1), \ldots, m(\varphi_n))$. From this definition it follows that $\neg$ is an unary logical function and $\vee$, $\wedge$ are binary logical functions.

Let us take some unary predicates $P_1, \ldots, P_k$ verified on the domain of D and let us take some variables $x, y, \ldots$ that are interpreted as members of D. Then we can extend $\mathfrak{I}$ to $\mathfrak{I}_p$ by adding (i) atomic formulas $P_1(x)$, $\ldots, P_k(x)$, (ii) their propositional superpositions constructed by using negation $\neg$, conjunction $\wedge$, and disjunction $\vee$; and (iii) quantified formulas $\forall x\varphi$ (to read 'φ for all x') and $\exists x\varphi$ (to read 'φ for some x'), where φ is an atomic formula $P(x)$ or a propositional superpositions of atomic formulas $P_1(x), \ldots, P_l(x)$. Then the logical meaning m can be expanded as follows:

- let x be a variable, then $m(x)$ means an element from D;
- let $P(x)$ be an atomic formula of $\mathfrak{I}_p$, then $m(P(x)) = \top$ if and only if $P(m(x)) \subseteq D$, otherwise $m(P(x)) = \bot$;
- let $f(\varphi_1, \ldots, \varphi_i)$ be a logical function defined on atomic formulas or their propositional superpositions $\varphi_1, \ldots, \varphi_i$ from $\mathfrak{I}_p$, then $m(f(\varphi_1, \ldots, \varphi_i)) = \top$ if and only if $f(m(\varphi_1), \ldots, m(\varphi_i)) = \top$, otherwise $f(m(\varphi_1), \ldots, m(\varphi_i)) = \bot$;
- let $\forall x\varphi$ be a quantified formula of $\mathfrak{I}_p$, then $m(\forall x\varphi) = \top$ if and only if for every element a in the domain of D, $m(x) = a$ and we have $m(\varphi) = \top$, otherwise $m(\forall x P(x)) = \bot$ (for instance, $m(\forall x(P_i(x) \wedge P_j(x)) = \top$ if and only if $P_i(a) \subseteq D$ and $P_j(a) \subseteq D$ for all $a \in D$);

- let $\exists x\varphi$ be a quantified formula of $\mathfrak{I}_P$, then $m(\exists x\varphi) = \top$ if and only if there is an element a in the domain of D, such that $m(x) = a$ and we have $m(\varphi) = \top$, otherwise $m(\exists x P(x)) = \bot$.

Now, let us take two unary modal operators: $\square$ and $\lozenge$, defined on formulas φ from $\mathfrak{I}_P$. Let us introduce the so-called modal formulas: $\square\varphi$ (to read: 'it is necessary that φ') and $\lozenge\varphi$ (to read: 'it is possible that φ'). The set $\mathfrak{I}_P$ is extended again to $\mathfrak{I}_\square$ by adding modal formulas. Assume that X is a set of indices and there is a distinguished index a. Let R be a binary relation on the indices. The distinguished index a is to represent an actual time. The relation R is to be said a possibility since a. Let us take x from X such that aRx. Then this aRx shows us a possibility since a at an index x. Then we can again extend m:

- for a formula $\varphi \in \mathfrak{I}_\square$ without modal operators, $m(\varphi) = \top$ at $x \in X$ if and only if $m(\varphi) = \top$;
- for a formula $\varphi \in \mathfrak{I}_\square$ without modal operators, $m(\varphi) = \bot$ at $x \in X$ if and only if $m(\varphi) = \bot$;
- let $\square\varphi$ be a modal formula of $\mathfrak{I}_\square$, then $m(\square\varphi) = \top$ at $x \in X$ if and only if for all $y \in X$ with xRy, $m(\varphi) = \top$ at y, otherwise $m(\square\varphi) = \bot$ at x;
- let $\lozenge\varphi$ be a modal formula of $\mathfrak{I}_\square$, then $m(\lozenge\varphi) = \top$ at $x \in X$ if and only if for some $y \in X$ with xRy, $m(\varphi) = \top$ at y, otherwise $m(\lozenge\varphi) = \bot$ at x;
- let $\forall x\varphi$ be a quantified formula of $\mathfrak{I}_\square$, then $m(\forall x\varphi) = \top$ at $t \in X$ if and only if for every element a in the domain of D, $m(x) = a$ and we have $m(\varphi) = \top$ at t, otherwise $m(\forall x P(x)) = \bot$ at t;
- let $\exists x\varphi$ be a quantified formula of $\mathfrak{I}_\square$, then $m(\exists x\varphi) = \top$ at $t \in X$ if and only if there is an element a in the domain of D, such that $m(x) = a$ and we have $m(\varphi) = \top$ at t, otherwise $m(\exists x P(x)) = \bot$ at t.

Thus, finally, an n-place *two-valued logical function* f is defined as a mapping from $\mathfrak{I}_\square^n$ to $\mathfrak{I}_\square$ such that $m(f(\varphi_1, \ldots, \varphi_n)) = f(m(\varphi_1), \ldots, m(\varphi_n))$. So, $\forall$ and $\exists$ as well as $\mathfrak{I}$ and $\lozenge$ are unary logical functions.

3. Logical Duality

Suppose that f is an n-place two-valued logical function. Another n-place two-valued logical function f' is said to be *dual* (or *logically dual*) to f if and only if

$$f'(\varphi_1, \ldots, \varphi_n) \equiv \neg f(\neg\varphi_1, \ldots, \neg\varphi_n),$$

where $\equiv$ is a sign for the equivalence relation: $f'(\varphi_1, \ldots, \varphi_n)$ is true if and only if $\neg f(\neg\varphi_1, \ldots, \neg\varphi_n)$ is true and $f'(\varphi_1, \ldots, \varphi_n)$ is false if and only if $\neg f(\neg\varphi_1, \ldots, \neg\varphi_n)$ is false.

According to this definition, if f' is dual to f, then f is dual to f'. So, the duality is always mutual.

Let us observe that conjunction and disjunction are dual to each other:

$$(\varphi \wedge \psi) \equiv \neg(\neg\varphi \vee \neg\psi);$$
$$(\varphi \vee \psi) \equiv \neg(\neg\varphi \wedge \neg\psi).$$

The universal quantifier $\forall$ and the existential quantifier $\exists$ are dual to each other, too:

$$\forall x\varphi \equiv \neg\exists x\neg\varphi;$$
$$\exists x\varphi \equiv \neg\forall x\neg\varphi.$$

The necessity modal operator $\square$ and the possibility modal operator $\lozenge$ are another example of logical duality:

$$\square\varphi \equiv \neg\lozenge\neg\varphi;$$
$$\lozenge\varphi \equiv \neg\square\neg\varphi.$$

The *logical duality* is a significant concept of logic, because it allows us to define a lattice, i.e. it can be used for ordering logical functions. Indeed, if f and f' are dual to each other, then either $(f \Rightarrow f') \equiv \top$ or $(f' \Rightarrow f) \equiv \top$, where $\Rightarrow$ is a sign for implication (a two-place two-valued logical function "if ..., then ...") such that

$$(\varphi \Rightarrow \psi) \equiv (\neg\varphi \vee \psi).$$

It means that either $m(f) \leq m(f')$ or $m(f') \leq m(f)$ by assuming that $\bot \leq \top$. So, we have the following true implications:

$$(\varphi \wedge \psi) \Rightarrow (\varphi \vee \psi);$$
$$\forall x\varphi \Rightarrow \exists x\varphi.$$

Let us check the first claim. Assume that $\top \equiv (\neg\varphi \vee \varphi)$. Then $((\varphi \wedge \psi) \Rightarrow (\varphi \vee \psi)) \equiv (\neg(\varphi \wedge \psi) \vee (\varphi \vee \psi)) \equiv (\neg\varphi \vee \neg\psi \vee \varphi \vee \psi) \equiv ((\neg\varphi \vee \varphi) \vee (\neg\psi \vee \psi)) \equiv \top \vee \top \equiv \top$. In the modal logic **D**, the true implication is as follows:

$$\square\varphi \Rightarrow \lozenge\varphi.$$

Based on logical duality, we can define contrary, subcontrary, subaltern, and contradictory logical functions:

contrary: two functions h and h' are *contrary* if and only if $(h \wedge h') \equiv \bot$, but not always $(h \vee h') \equiv \top$;

subcontrary: two functions h and h' are *subcontrary* if and only if $(h \vee h') \equiv \top$, but not always $(h \wedge h') \equiv \bot$;

subaltern: a function h is *subaltern* to h' if and only if $(h' \Rightarrow h) \equiv \top$;

contradictory: two functions h and h' are *contradictory* if and only if $(h \vee h') \equiv \top$ and $(h \wedge h') \equiv \bot$.

Let us assume that $f(\varphi_1, \ldots, \varphi_n)$ and $f'(\varphi_1, \ldots, \varphi_n)$ are dual and $(f(\varphi_1, \ldots, \varphi_n) \Rightarrow f'(\varphi_1, \ldots, \varphi_n)) \equiv \top$. Then

- $f'(\varphi_1, \ldots, \varphi_n)$ is *subaltern* to $f(\varphi_1, \ldots, \varphi_n)$;
- $f(\varphi_1, \ldots, \varphi_n)$ and $f(\neg\varphi_1, \ldots, \neg\varphi_n)$ are *contrary*;
- $f'(\varphi_1, \ldots, \varphi_n)$ and $f'(\neg\varphi_1, \ldots, \neg\varphi_n)$ are *subcontrary*;
- $f'(\varphi_1, \ldots, \varphi_n)$ and $f(\neg\varphi_1, \ldots, \neg\varphi_n)$ (as well as $f(\varphi_1, \ldots, \varphi_n)$ and $f'(\neg\varphi_1, \ldots, \neg\varphi_n)$) are *contradictory*.

Contrary, subcontrary, subaltern, and contradictory relations for the following four logical functions: $f(\varphi_1, \ldots, \varphi_n)$; $f'(\varphi_1, \ldots, \varphi_n)$; $f(\neg\varphi_1, \ldots, \neg\varphi_n)$; $f'(\neg\varphi_1, \ldots, \neg\varphi_n)$, are called a *logical square of opposition*, please see Fig. 1. So, we can obtain a logical square of opposition for the following four logical functions: $(p \wedge q)$; $(\neg p \wedge \neg q)$; $(p \vee q)$; $(\neg p \vee \neg q)$, see Fig. 2. In the same way, there is a square of opposition for the following atomic propositions: $\forall x P(x)$; $\forall x \neg P(x)$; $\exists x P(x)$; $\exists x \neg P(x)$, see Fig. 3, and for the following modal formulas: $\square p$; $\square \neg p$; $\lozenge p$; $\lozenge \neg p$, see Fig. 4.

The logical square of opposition exemplified from Figs. 1 to 4 can have different modifications considered in [3] and [4]. So, we can introduce a *logical cube of opposition* if we focus on binary logical functions $f(\varphi_1, \varphi_2)$; $f(\varphi_1, \neg\varphi_2)$; $f(\neg\varphi_1, \varphi_2)$; $f(\neg\varphi_1, \neg\varphi_2)$; $f'(\varphi_1, \varphi_2)$; $f'(\varphi_1, \neg\varphi_2)$; $f'(\neg\varphi_1, \varphi_2)$; $f'(\neg\varphi_1, \neg\varphi_2)$, where $f'(\varphi_1, \varphi_2)$ is subaltern to $f(\varphi_1, \varphi_2)$ (it means that they are dual logical functions), $f'(\varphi_1, \neg\varphi_2)$ is subaltern to $f(\varphi_1, \neg\varphi_2)$, $f'(\neg\varphi_1, \varphi_2)$ is subaltern to $f(\neg\varphi_1, \varphi_2)$, and $f'(\neg\varphi_1, \neg\varphi_2)$ is subaltern to $f(\neg\varphi_1, \neg\varphi_2)$. An example of cube of opposition for $\vee$ and $\wedge$ is given in Fig. 5. Also, the cube of opposition can be defined for compositions of two unary logical functions $\forall \circ \square$, $\forall \circ \lozenge$, $\exists \circ \square$, $\exists \circ \lozenge$, see Fig. 6.

Hence, the logical duality allows us to define an order over propositional formulas. It is a necessary step before constructing a formal logical system. For more details on duality, please see [3 and 4].

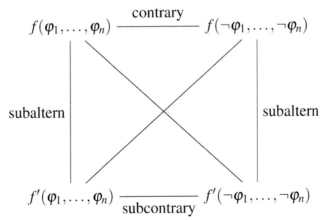

Fig. 1. The square of opposition for the dual logical functions
$f(\varphi_1, \ldots, \varphi_n)$ and $f'(\varphi_1, \ldots, \varphi_n)$.

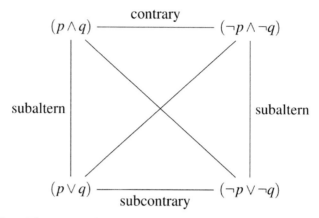

Fig. 2. The square of opposition for the binary operators ∧ and ∨.

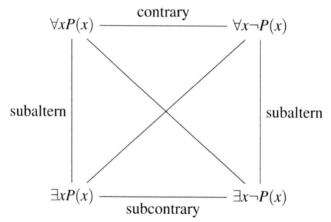

Fig. 3. The square of opposition for the quantifiers ∃ and ∀.

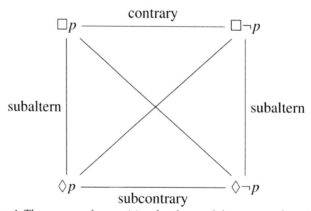

Fig. 4. The square of opposition for the modal operators ◊ and □.

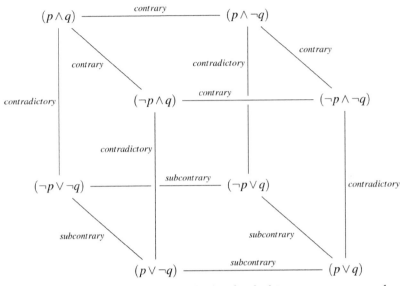

Fig. 5. The cube of generalized Post duality for the binary operators ∧ and ∨.

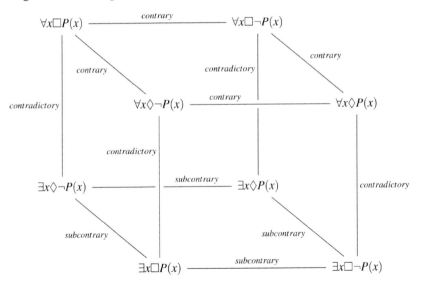

Fig. 6. Duality cube for the following composed operators:
∀ ∘ □; ∀ ∘ ◊; ∃ ∘ □; ∃ ∘ ◊.

4. Logical Interpretation of Swarm Reactions

Each swarm may be regarded as a *multi-agent system* consisting of n actors (members of this swarm) $A_1, \ldots, A_n$. All the behaviors of the actors are influenced by biologically active things of two types, see [5] and [6]: (i) *repellents* (such as hazardous substances or predators), which are always avoided by the actors, and (ii) *attractants* (such as food or sexual pheromone) which attract them.

Each member of swarm can detect a repellent or attractant in a radius not longer than r. Hence, let us assume that a repellent or attractant X is placed at the distance $d \leq r$ from an agent A. It means that this X, respectively, repels or attracts the swarm member A. This property allows us to interpret propositional variables $p \in \mathfrak{I}_\square$ as biological stimuli: (i) $m_{X,t}^A(p) = \top$ for the agent A and a biologically active thing X at the time step t if and only if A is attracted by X at t; (ii) $m_{X,t}^A(p) = \bot$ for the agent A and a biologically active thing X at the time step t if and only if A is repelled by X at t.

Let us interpret the negation $\neg p$ of propositional variables $p \in \mathfrak{I}_\square$ as a biological stimulus:

true negation: $m_{X,t}^A(\neg p) = \top$ for the agent A and biologically active element X at the time step t if and only if $\neg(m_{X,t}^A(p)) = \top$, i.e. X is not occupied by A at t;

false negation: $m_{X,t}^A(\neg p) = \bot$ for the agent A and biologically active element X at the time step t if and only if $\neg(m_{X,t}^A(p)) = \bot$, i.e. X is not avoided by A at t.

Now, let us assume the existence of k agents $A_1, \ldots, A_k$ with the radii of sensitivity $rA_1, \ldots, rA_k$, respectively. Suppose that there are two different active things X_1 and X_2 located at the distance d such that $d \leq rA_1, \ldots, d \leq rA_k$. There are two possibilities, provided that X_1 and X_2 are attractants: (i) both X_1 **and** X_2 are occupied by the members $A_1, \ldots, A_k$ (for instance, X_1 is occupied by A_1 and X_2 is occupied by A_2, or X_1 is occupied by A_2 and X_2 is occupied by A_1, etc.); (ii) X_1 **or** X_2 are occupied by the members $A_1, \ldots, A_k$ (for instance, X_1 is occupied by A_1 and A_2 simultaneously, or X_2 is occupied by A_1 and A_2 simultaneously, etc.). Also, there are two possibilities, provided that X_1 and X_2 are repellents: (iii) both X_1 **and** X_2 are avoided by the members $A_1, \ldots, A_k$ (e.g. X_1 is avoided by A_1 and X_2 is avoided by A_2, or X_1 is avoided by A_2 and X_2 is avoided by A_1, etc.); (iv) X_1 **or** X_2 are avoided by the members $A_1, \ldots, A_k$ (e.g. X_1 is avoided by A_1 and A_2 simultaneously, or X_2 is avoided by A_1 and A_2 simultaneously, etc.). In the third situation that one of X_1 and X_2 is a repellent and another active element is an attractant, we face the following possibilities: (v) both X_1 **and** X_2 affect the members $A_1, \ldots, A_k$; (iv) X_1 **or** X_2 affect the members $A_1, \ldots, A_k$.

Thus, we can interpret the conjunction $p \wedge q$ of propositional variables p, q as a biological stimulus, too:

true conjunction: $m_{X_1, X_2, t}^{A_1, \ldots, A_k} (p \wedge q) = \top$ for the agents $A_1, \ldots, A_k$ and biologically active elements X_1 and X_2 at the time step t if and only if $(m_{X_1, X_2, t}^{A_1, \ldots, A_k} (p) \wedge m_{X_1, X_2, t}^{A_1, \ldots, A_k} (q)) = \top$, i.e. both X_1 and X_2 are occupied by $A_1, \ldots, A_k$ at t;

false conjunction: $m_{X_1, X_2, t}^{A_1, \ldots, A_k} (p \wedge q) = \bot$ for the agents $A_1, \ldots, A_k$ and biologically active elements X_1 and X_2 at the time step t if and only if $(m_{X_1, X_2, t}^{A_1, \ldots, A_k} (p) \wedge m_{X_1, X_2, t}^{A_1, \ldots, A_k} (q)) = \bot$, i.e. X_1 or X_2 are avoided by $A_1, \ldots, A_k$ at t.

The same concerns the disjunction $p \vee q$ of propositional variables $p, q \in \mathfrak{I}_\square$:

true disjunction: $m_{X_1, X_2, t}^{A_1, \ldots, A_k} (p \vee q) = \top$ for the agents $A_1, \ldots, A_k$ and biologically active elements X_1 and X_2 at the time step t if and only if $(m_{X_1, X_2, t}^{A_1, \ldots, A_k} (p) \vee m_{X_1, X_2, t}^{A_1, \ldots, A_k} (q)) = \top$, i.e. X_1 or X_2 are occupied by $A_1, \ldots, A_k$ at t;

false disjunction: $m_{X_1, X_2, t}^{A_1, \ldots, A_k} (p \vee q) = \bot$ for the agents $A_1, \ldots, A_k$ and biologically active elements X_1 and X_2 at the time step t if and only if $(m_{X_1, X_2, t}^{A_1, \ldots, A_k} (p) \vee m_{X_1, X_2, t}^{A_1, \ldots, A_k} (q)) = \bot$, i.e. both X_1 and X_2 are avoided by $A_1, \ldots, A_k$ at t.

Let l be a maximal number of different active things which can be detected at the time t by all the members $A_1, \ldots, A_k$. Then we can generalize our new semantics as follows:

true l-place logical function: $m_{X_1, \ldots, X_l, t}^{A_1, \ldots, A_k} (f(p_1, \ldots, p_l) = \top$ for the agents $A_1, \ldots, A_k$ and biologically active elements $X_1, \ldots, X_l$ at the time step t if and only if $f(m_{X_1, \ldots, X_l, t}^{A_1, \ldots, A_k} (p_1), \ldots, m_{X_1, \ldots, X_l, t}^{A_1, \ldots, A_k} (p_l)) = \top$;

false l-place logical function: $m_{X_1, \ldots, X_l, t}^{A_1, \ldots, A_k} (f(p_1, \ldots, p_l) = \bot$ for the agents $A_1, \ldots, A_k$ and biologically active elements $X_1, \ldots, X_l$ at the time step t if and only if $f(m_{X_1, \ldots, X_l, t}^{A_1, \ldots, A_k} (p_1), \ldots, m_{X_1, \ldots, X_l, t}^{A_1, \ldots, A_k} (p_l)) = \bot$.

Hence, in our definition of realizing l-place logical function on the swarm behavior, we assume a bijection between the set of propositional variables and the set of active elements affecting at the time step t.

Let us show that $m_{X, t}^{A_1, \ldots, A_k} (p \vee \neg p) = \top$. Indeed, $m_{X, t}^{A_1, \ldots, A_k} (p \vee \neg p) = (m_{X, t}^{A_1, \ldots, A_k} (p) \vee \neg m_{X, t}^{A_1, \ldots, A_k} (p)) = \top$. It means that the active element X, reachable for $A_1, \ldots, A_k$, either attract $A_1, \ldots, A_k$ or repel $A_1, \ldots, A_k$. On the other hand, $m_{X, t}^{A_1, \ldots, A_k} (p \wedge \neg p) = \bot$. So, $m_{X, t}^{A_1, \ldots, A_k} (p \wedge \neg p) = (m_{X, t}^{A_1, \ldots, A_k} (p) \wedge \neg m_{X, t}^{A_1, \ldots, A_k} (p)) = \bot$, because the same active element X, reachable for $A_1, \ldots, A_k$, cannot attract and repel $A_1, \ldots, A_k$ at the same time.

Also, we can behaviorally interpret unary predicates $P_1, \ldots, P_k$ verified on the domain of D as well as variables $x, y, \ldots$ understood as members of D. Let D_t mean all the biologically active elements reachable for $A_1, \ldots, A_k$ at the time t. Then $D = \bigcup_{t=0}^{n} D_t$, where n is a time step denoting one life cycle of a swarm.

True atomic formula: $m_{D,t}^{A_1, \ldots, A_k}(P(x)) = \top$ for $A_1, \ldots, A_k$ and biologically active elements of D at the time step t if and only if $P(m_{D,t}^{A_1, \ldots, A_k}(x)) \subseteq D_t$, where $m_{D,t}^{A_1, \ldots, A_k}(x)$ means an element from D_t. For instance, let P mean 'neighbors for the attractant X (i.e. elements reachable from X at one step of agents $A_1, \ldots, A_k)'$. Then $m_{D,t}^{A_1, \ldots, A_k}(P(x)) = \top$ if and only if there are neighbors for X in D at t.

False atomic formula: $m_{D,t}^{A_1, \ldots, A_k}(P(x)) = \bot$ for $A_1, \ldots, A_k$ and biologically active elements of D at the time step t if and only if $P(m_{D,t}^{A_1, \ldots, A_k}(x)) \not\subseteq D_t$, where $m_{D,t}^{A_1, \ldots, A_k}(x)$ means an element from D_t.

Let $f(\varphi_1, \ldots, \varphi_i)$ be a logical function defined on atomic formulas or their propositional superpositions $\varphi_1, \ldots, \varphi_i$ from $\mathfrak{I}_p$, then $m_{D,t}^{A_1, \ldots, A_k}(f(\varphi_1, \ldots, \varphi_i)) = \top$ if and only if $f(m_{D,t}^{A_1, \ldots, A_k}(\varphi_1), \ldots, m_{D,t}^{A_1, \ldots, A_k}(\varphi_i)) = \top$, otherwise $f(m_{D,t}^{A_1, \ldots, A_k}(\varphi_1), \ldots, m_{D,t}^{A_1, \ldots, A_k}(\varphi_i)) = \bot$.

Suppose, $\forall x \varphi$ is a quantified formula of $\mathfrak{I}_p$, then $m_{D,t}^{A_1, \ldots, A_k}(\forall x \varphi) = \top$ if and only if for every element a in the domain of D_t, $m_{D,t}^{A_1, \ldots, A_k}(x) = a$ and we have $m_{D,t}^{A_1, \ldots, A_k}(\varphi) = \top$, otherwise $m_{D,t}^{A_1, \ldots, A_k}(\forall x P(x)) = \bot$. Let $\exists x \varphi$ be a quantified formula of $\mathfrak{I}_p$, then $m_{D,t}^{A_1, \ldots, A_k}(\exists x \varphi) = \top$ if and only if there is an element a in the domain of D_t, such that $m_{D,t}^{A_1, \ldots, A_k}(x) = a$ and we have $m_{D,t}^{A_1, \ldots, A_k}(\varphi) = \top$, otherwise $m_{D,t}^{A_1, \ldots, A_k}(\exists x P(x)) = \bot$.

Modal formulas of $\mathfrak{I}_\square$ can be behaviorally interpreted, too. Let the index set X consist of indices denoting different time steps within one life cycle of a swarm.

- for a formula $\varphi \in \mathfrak{I}_\square$ without modal operators, $m_{\square,t}^{A_1, \ldots, A_k}(\varphi) = \top$ at $t \in X$ if and only if $m_{D,t}^{A_1, \ldots, A_k}(\varphi) = \top$ or $m_{X_1, \ldots, X_l t}^{A_1, \ldots, A_k}(\varphi) = \top$;

- for a formula $\varphi \in \mathfrak{I}_\square$ without modal operators, $m_{\square,t}^{A_1, \ldots, A_k}(\varphi) = \bot$ at $t \in X$ if and only if $m_{D,t}^{A_1, \ldots, A_k}(\varphi) = \bot$ or $m_{X_1, \ldots, X_l t}^{A_1, \ldots, A_k}(\varphi) = \bot$;

- let $\square \varphi$ be a modal formula of $\mathfrak{I}_\square$, then $m_{\square,t}^{A_1, \ldots, A_k}(\square\varphi) = \top$ at $x \in X$ if and only if for all $y \in X$ with xRy, $m_{\square,t}^{A_1, \ldots, A_k}(\varphi) = \top$ at y, otherwise $m_{\square,t}^{A_1, \ldots, A_k}(\square\varphi) = \bot$ at x;

- let $\lozenge\varphi$ be a modal formula of $\mathfrak{I}_\square$, then $m_{\square,t}^{A_1, \ldots, A_k}(\lozenge\varphi) = \top$ at $x \in X$ if and only if for some $y \in X$ with xRy, $m_{\square,t}^{A_1, \ldots, A_k}(\varphi) = \top$ at y, otherwise $m_{\square,t}^{A_1, \ldots, A_k}(\lozenge\varphi) = \bot$ at x;

- let $\forall x \varphi$ be a quantified formula of $\mathfrak{I}_\square$, then $m_{\square,t}^{A_1, \ldots, A_k}(\forall x \varphi) = \top$ at $t \in X$ if and only if for every element a in the domain of D_t, $m_{\square,t}^{A_1, \ldots, A_k}(x) = a$

and we have $m_{\square,t}^{A_1,...,A_k}(\varphi) = \top$ at t, otherwise $m_{\square,t}^{A_1,...,A_k}(\forall x P(x)) = \bot$ at t;

- let $\exists x\varphi$ be a quantified formula of $\Im_\square$, then $m_{\square,t}^{A_1,...,A_k}(\exists x\varphi) = \top$ at $t \in X$ if and only if there is an element a in the domain of D_t, such that $m_{\square,t}^{A_1,...,A_k}(x) = a$ and we have $m_{\square,t}^{A_1,...,A_k}(\varphi) = \top$ at t, otherwise $m_{\square,t}^{A_1,...,A_k}(\exists x P(x)) = \bot$ at t.

Hence, all the formulas of $\Im_\square$ can be verified on different reactions of swarm. In this way, swarms realize the logical duality of Figs. 1–6.

5. Logical Duality in Swarm Reactions

We have just defined the logical negation on the natural swarm behavior as the avoidance of a place by the swarm members. But it is still unknown, why the swarm members prefer to realize either the logical disjunction or logical conjunction under different conditions. Nevertheless, there is a natural explanation of both types of these swarm reactions. So, the swarm members can be either under stress or with a sense of safety. For example, if some swarm individuals face two attractants, they experience a feeling of safety and if they face two repellents, they experience stress. Also, a stress can be caused by a very high concentration of attractive pheromone. In any case, there is a natural duality in the swarm reactions: these reactions are carried out either under stress or with sense of safety, see [5].

Let us define two unary predicates on propositional variables p: 'stress from p' and 'safety from p'. The variables p are interpreted as biologically active elements (attractants or repellents). Standardly, an attractant with a usual concentration of pheromone causes safety and a repellent with a usual concentration of dangerous matter causes stress. Let $\neg p$ be interpreted as a complement to the active element of p in the class of all active elements. Thus, we can construct a square of opposition, see Fig. 7.

In this square of opposition, the predicates 'stress from p' and 'safety from $\neg p$' are considered dual: if 'stress from p' holds true, then 'safety from $\neg p$' holds true. However, it is possible to claim that the feeling of stress and the feeling of safety are two psychological phenomena which concern individuals of birds and mammals and they do not concern insects or unicellular organisms surely. But then how can we explain the same swarm reactions of social insects and social bacteria? They also prefer either conjunction or disjunction under different conditions. The point is that 'stress from p' and 'safety from p' can be realized by an individual even such as an insect or bacterium without any mechanism of awareness.

In each networking, including even networks of actin filaments in one cell, there are two basic reactions to outer stimuli: *lateral activation* (a reaction under safety) and *lateral inhibition* (a reaction under stress).

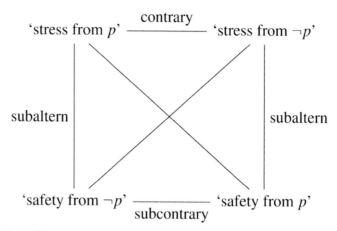

Fig. 7. The square of opposition for the expressions 'relax from p' and 'stress from $\neg p$'.

Both types of reactions can be realized without awareness and explained rather chemically [5]. For instance, these types of reactions are observed in plasmodia of *Physarum polycephalum* – an amoeboid multinucleated organism. It is evident that this organism does not have feelings at all.

The *lateral activation* is a reaction of particles within one network to outer stimuli, according to which different particles are not concentrated on the same stimuli. As a result, we observe a decreasing of the intensity of the outer signals and the contrast of the signals is made less visible. The *lateral inhibition* is a reaction of particles within one network to outer stimuli, according to which different particles are concentrated on the same stimuli. This has led us to an increasing of the intensity of the outer signals and the contrast of the signals is made more visible. The plasmodia of *Physarum polycephalum* follow the lateral activation if they detect normal attractants and they follow the lateral inhibition if they face standard repellents [5].

Thus, let us define two unary operators: (i) $INH_{A_1,\ldots,A_k}(p)$ ::= 'the swarm individuals $A_1, \ldots, A_k$ are laterally inhibited by an outer signal p'; $INH_{A_1,\ldots,A_k}(p) \equiv \top$ if and only if $(\neg p) \equiv \top$; (ii) $ACT_{A_1,\ldots,A_k}(p)$ ::= 'the swarm individuals $A_1, \ldots, A_k$ are laterally activated by an outer signal p'; $ACT_{A_1,\ldots,A_k}(p) \equiv \top$ if and only if $(\neg p) \equiv \bot$. There is a logical duality between $INH_{A_1,\ldots,A_k}(p)$ and $ACT_{A_1,\ldots,A_k}(\neg p)$:

$$(INH_{A_1,\ldots,A_k}(p) \Rightarrow ACT_{A_1,\ldots,A_k}(\neg p)) \equiv \top.$$

As a consequence, we can introduce a logical square of opposition for $INH_{A_1,\ldots,A_k}(p)$ and $ACT_{A_1,\ldots,A_k}(\neg p)$, please see Fig. 8. It is the same as in Fig. 7, but it is defined more correctly from the point of view of cognitive science, because 'stress' and 'safety' are not precise terms.

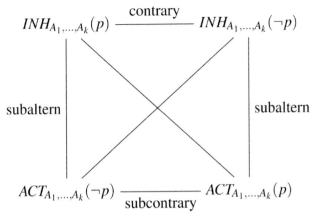

Fig. 8. The square of opposition for the unary operators $INH_{A_1,...,A_k}$ and $ACT_{A_1,...,A_k}$.

Now, we can define binary operators: (i) $INH_{A_1,...,A_k}(p, q)$ ::= 'the swarm individuals $A_1, ..., A_k$ are laterally inhibited by two outer signals p and q'; $INH_{A_1,...,A_k}(p, q) \equiv \top$ if and only if $(p \wedge q) \equiv \top$; (ii) $ACT_{A_1,...,A_k}(p, q)$ ::= 'the swarm individuals $A_1, ..., A_k$ are laterally activated by two outer signals p and q'; $ACT_{A_1,...,A_k}(p, q) \equiv \top$ if and only if $(\neg(p \wedge q)) \equiv (\neg p \vee \neg q) \equiv \top$. From these definitions it follows that

$$(INH_{A_1,...,A_k}(p, q) \Rightarrow ACT_{A_1,...,A_k}(\neg p, \neg q)) \equiv \top.$$

Hence, we obtain a square of opposition for $INH_{A_1,...,A_k}(p, q)$ and $ACT_{A_1,...,A_k}(\neg p, \neg q)$, see Fig. 9. This square has the same meaning as the square of Fig. 2.

Also, we can generalize the cube of opposition of Fig. 5 up to the duality on the swarm behavior. In this way, we obtain a cube of opposition for the binary operators $INH_{A_1,...,A_k}$ and $ACT_{A_1,...,A_k}$, see Fig. 10.

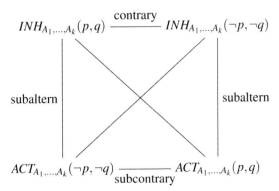

Fig. 9. The square of opposition for the binary operators $INH_{A_1,...,A_k}$ and $ACT_{A_1,...,A_k}$.

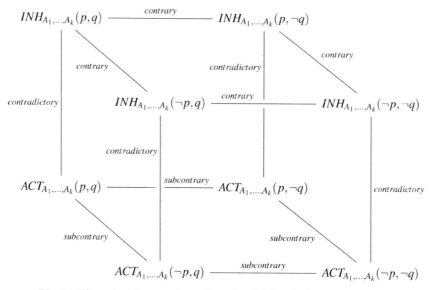

Fig. 10. The cube of generalized Post duality for the binary operators $INH_{A_1,...,A_k}$ and $ACT_{A_1,...,A_k}$.

To sum up, we see that swarms realize a kind of logical duality in their reactions towards outer stimuli p and q, since either they behave under lateral activation and realize the false conjunction of p and q or they can behave under lateral inhibition and realize the true conjunction of p and q, see Figs. 9 and 10.

6. Conclusion and Discussion

We have just considered an abstract model of swarms, where we have defined logical functions of the $\mathfrak{I}_\Box$ as natural reactions of swarm members to several stimuli detected at one time step. In this model, the logical duality represented by squares or cubes of opposition is realizable by their own swarm patterns based on lateral inhibition and lateral activation. Nevertheless, this model is abstract in fact. The problem is that swarm reactions are too sensitive to outer signals and these reactions assume that the number of inputs might be smaller than the number of outputs, see [6]. It means that swarms implement not standard logical functions, but their vague modifications. So, swarm reactions are not certain. We cannot say that they are either laterally inhibited or laterally activated. They can be partly inhibited and partly activated. They can be less or more inhibited and less or more activated, etc. So, we can deal with a fuzzy mix of conjunction and disjunction at one time. And this mix can have continuous modifications. So, *standard logical functions are just an idealization of real swarm reactions.* They are a digital representation of continuous mixes of different functions in one complex response to several outer stimuli.

References

1. A. Adamatzky, V. Erokhin, M. Grube, T. Schubert and A. Schumann. Physarum chip project: Growing computers from slime mould. International Journal of Unconventional Computing 8(4), 319–323 (2012).
2. C.A. Coello, R.L.G. Zavala, B.M. Garca and A. Hernndez Aguirre. Ant Colony System for the Design of Combinational Logic Circuits. *In*: J. Miller, A. Thompson, P. Thomson and T.C. Fogarty (eds), Evolvable Systems: From Biology to Hardware. ICES 2000. Lecture Notes in Computer Science, vol 1801. Springer, Berlin, Heidelberg, pp. 21–30 (2000).
3. L. Demey and H. Smessaert. Aristotelian and Duality Relations Beyond the Square of Opposition. *In*: Chapman, P., Stapleton, G., Moktefi, A., Perez-Kriz, S. and Bellucci F. (eds), Diagrammatic Representation and Inference. Diagrams, 2018. Lecture Notes in Computer Science, vol 10871. Springer, Cham, 2018, pp. 640–656.
4. L. Demey and H. Smessaert. Duality in Logic and Language. *In*: Internet Encyclopedia of Philosophy. https://www.iep.utm.edu/dual-log/
5. A. Schumann. Behaviourism in Studying Swarms: Logical Models of Sensing and Motoring. Emergence, Complexity and Computation, vol 33. Springer, Cham (2019).
6. A. Schumann and K. Pancerz. High-Level Models of Unconventional Computations: A Case of Plasmodium. Springer International Publishing (2019).

9

On the Motion of Agents with Directional Antennae

Alexander Kuznetsov

University of Information Technology and Management in Rzeszow, Poland
Email: avkuz@bk.ru

1. Introduction

In this chapter, I will discuss one of possible methods for organization of the formation control and connectivity management for agents like unmanned aerial vehicle (UAV). More precisely, I will propose a method of organizing a system of agents that have directional antennae and, therefore, have significant restrictions on the exchange of information about their location and the maintenance of a telecommunication network. This idea has already been briefly described in a previous conference paper [14].

The work [1] contains a study using of the directional antennae for UAV organization. It had found that the utilization of a higher gain directional antenna versus the standard omni-directional antenna yields more than twice the UAV communications range. They also had discovered that the range increase was a factor of antenna gain and propagation pattern and not change in a receiver/transmitter configuration. It was also obvious that the use of directional antennas increases the stealth of the agent communication system, which makes sense for use in military applications.

In the process of movement, such agents must continuously change the direction of their antennas to ensure continuous communication. If there is a base station in the agent network, it is possible to correct the antenna position from the base station. However, in a ad hoc network of agents, it is necessary to use algorithms similar to those used for organizing the structure of agents in decentralized systems.

In this case, the agent sends a message to the nearest neighbour about the need to change locations and further such information is distributed along a chain of neighbouring agents. In the case of a system of agents with directional antennas, it is necessary to transmit a message about the need to change the orientation of the antenna in order to maintain connectivity to the network.

In short, the idea of the algorithm for the movement and communication of such agents is as follows:

- All agents are at their initial positions, the antennas are directed so that the necessary connectivity of the agent network is achieved. Each agent has a list L of agents with whom he must keep in touch.
- If one of the agents needs a turn, the agent sends to the nearest agent ag_N from L message about the need to deploy the antenna for a given number of degrees.
- After receiving confirmation from ag_N, the agent makes the necessary turns to the antennas.
- The agent waits for the duration of the time required for all agents to turn the antennas from L.
- Agent continues to move.

Obviously, agents connected by directional antennas cannot move in a completely arbitrary manner and must also maintain a certain order, although not necessarily rigid. Rather, such a system can be described using the fuzzy graph previously introduced by the author of [13]. The relationship between the degree of rigidity of the graph of the system of agents and the angle of the solution of the main lobe of the antenna pattern of each agent is the subject of ongoing research by the author.

From the foregoing, it follows that each agent must be able to predict the direction of movement of other agents. Prediction occurs based on the latest obtained data on the direction and speed of the movement of the agents, i.e. it is assumed that for some time the direction of movement and the speed of the agent will be maintained.

We will try to apply stability based control for this kind of UAVs system and to overcome difficulties which follow from the communication system variability.

2. Stability Theory based Distributed Target-centric Formation Control

Several researchers [17], [9], [16], [3] have investigated the problem of designing formation control laws for connected networks. These control laws work effectively as long as the graph is connected [17], [16]. However, they do not take care of the connectivity of the time-varying graph, and they cannot guarantee to maintain the connectivity during an entire mission and can be inadequate if the graph breaks during the dynamic process. Sharma et al. [12] considered a constant adjacency matrix with binary elements, which does not capture the time-varying information sharing among the agents subjected to the formation control. Maintaining a connectivity of a mobile network is a challenging area of research [6], [19], [18]. Different researchers had developed potential field based control laws [6] (which are useful also for flocking a large amount of agents) and super-gradient optimization techniques [19] to maintain and control the dynamic connectivity in distributed networks. Obviously, these methods have their own limitations and suitable for specific scenarios; for example, the potential field approach has difficulty working with the asymmetric formation arrangement.

The papers [4, 5, 12, 17] present distributed target-centric formation control strategy for multiple unmanned aerial vehicles (UAVs) using Lyapunov's stability [10] and graph theories with an emphasis on consensus and cooperation. The formation is maintained around a target using a combination of a consensus protocol and a sliding mode control law.

Consensus helps in distributing the target information which is available only to a subset of vehicles. The authors have employed a sliding mode controller to achieve a target centered formation whose information is partially available to the capturing vehicles. The connectivity of a network of unmanned systems changes as the state dependent graph evolves over time, revealing the risk of the system being uncontrollable during null connectivity.

Dutta, Kothari et al. [4] shows that a target-centric formation can be maintained, if at least one vehicle in a group has target information with some uncertainty, and the corresponding communication graph is connected.

Denote as R in the communication range,

$$V = \{a_1, \dots, a_{n-1}, a_n\} \tag{1}$$

is the set of all nodes (which are agents) of the communication graph, and p_i is a position of the node a_i. In this work, we assume that $a_i = i$, $i = \overline{1, n}$. The node a_n is reserved for the "target", i.e. an central agent for the formation setting. Put also $r_{ij} = p_i - p_j$.

Define the neighbourhood of the agent a_i as

$$N_i = \{j \in V, j \neq i: \| r_{ij} \| \leq R\}.$$

We will denote the adjacency matrix or the weighted adjacency matrix of the graph as A. In the case of the adjacency matrix, $a_{ij} = 1$ if agents a_i and a_j can mutually communicate and $a_{ij} = 0$ otherwise. In the case of the weighted adjacency matrix, $a_{ij} \geq 0$ is a quality of the link between a_i and a_j, a_{ij} is the worst quality. For example,

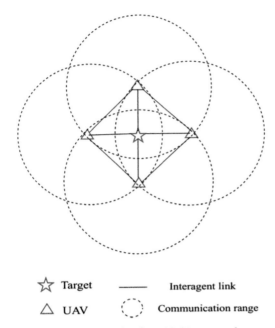

☆ Target   ———   Interagent link

△ UAV   ⟨ ⟩   Communication range

Fig. 1. Desired formation for four UAVs around one target (from the article [4]).

$$a_{ij} = e^{-\sigma \|r_{ij}\|},$$

where $\sigma > 0$ is a constant depending on R representing the slope of communication quality over distance. This dependence on the distance is conditioned by physical reasons: the sensors' resolution is decaying exponentially with the distance to the object to observe.

The degree matrix of a graph is the diagonal matrix D with elements on the main diagonal

$$d_{ii} = \sum_{j=1}^{n} a_{ij}.$$

Define elements l_{ij} of the Laplace $L = D - A$ matrix as

$$l_{ij} = \begin{cases} -a_{ij}, i \neq j; \\ \sum_{i=1, i \neq j}^{n} a_{ij}. \end{cases}$$

The second smallest eigenvalue $\lambda_2 \geq 0$ of the Laplace matrix is called the algebraic connectivity of the graph, and the associated eigenvector is known as the Fiedler vector. A higher positive value of λ_2 indicates a better connected graph. If this value goes down to zero, then the graph is not connected. In other words, there exists no spanning tree [8] in the graph.

The motion of the ith UAV is governed by the double-integrator model

$$\dot{p} = \begin{bmatrix} \dot{x}_i \\ \dot{y}_i \end{bmatrix} = \begin{bmatrix} v_i \cos \varphi_i \\ v_i \sin \varphi_i \end{bmatrix} \tag{2}$$

$$\ddot{p}_i = \begin{bmatrix} \cos \varphi_i & -v_i \sin \varphi_i \\ \sin \varphi_i & v_i \cos \varphi_i \end{bmatrix} \begin{bmatrix} \dot{v}_i \\ \dot{\varphi}_i \end{bmatrix} \tag{3}$$

$$\ddot{p}_i = M_i(\dot{p}_i)u_i; u_i = \begin{bmatrix} \dot{v}_i \\ \dot{\varphi}_i \end{bmatrix}; M_i = \begin{bmatrix} \cos \varphi_i & -v_i \sin \varphi_i \\ \sin \varphi_i & v_i \cos \varphi_i \end{bmatrix}. \tag{4}$$

The objective of a target-centric formation is to drive the UAVs at a constant distance δ and angle ψ_i with respect to the target. Let, the position and the velocity of the target be denoted by $e = p_n$ and $v = v_n$. The control objective can be expressed mathematically as follows

$$p_i(t) - e(t) \rightarrow P_i \tag{5}$$

$$\dot{p}_i(t) - \dot{e}_t(t) \rightarrow 0 \tag{6}$$

$$\psi_{i+1} - \psi_i = 2\pi/n \tag{7}$$

$$\lambda_2(L(t)) > 0; \tag{8}$$

where $P_i = \delta[\cos \psi_i, \sin \psi_i]^T$ for some pre-defined δ. The velocities of agents are bounded by $v_{min} \leq v_i \leq v_{max}$.

In control theory, backstepping is a technique developed circa 1990 by Petar V. Kokotovic [11] and others for designing stabilizing controls for a special class of nonlinear dynamical systems such as

$$\dot{\eta} = f(\eta) + g(\eta)\xi,$$

$$\dot{\xi} = u,$$

where $[\eta, \xi]^T \in \mathbb{R}^{n+1}$ is the state and $u \in \mathbb{R}$ is the control input.

Let us denote $\hat{P}_i = p_i - P_i$. Next we define new state variables for the two vehicles:

$$\tilde{r}_{ij} = \hat{P}_i - \hat{P}_j,$$

$$v_{ij} = \dot{\tilde{r}}_{ij} = \dot{p}_i - \dot{p}_j.$$

Using these variables, we obtain

$$\dot{\tilde{r}}_{ij} = v_{ij}, \tag{9}$$

$$\dot{v}_{ij} = M_i(v_{ij} + v_{ji})u_i - M_j(v_{ij} - v_{ji})u_j. \tag{10}$$

Furthermore, the works describe various options for creating a controller u_i which forces

$$\lim_{t \rightarrow \infty} r_{ij}(t) = 0.$$

If we assume

$$M_i u_i = M_j u_j - \alpha_1(\hat{p}_i - \hat{p}_j) - \alpha_2(\dot{p}_i - \dot{p}_j),$$

$\alpha_1 > 0$, $\alpha_2 > 0$, we obtain that the origin is globally asymptotically stable.

The control u_i should satisfy the n equalities. However, there is no u_i which satisfies all these equalities together and the system is overdetermined.

It is possible to suggest the solution which is the average of the u_i that satisfy these n equalities separately.

Also, the alternative variant can be studied. Introduce the following block vectors and matrices

$$\mathbf{r} = \begin{bmatrix} \tilde{r}_{1,n} \\ \cdots \\ \tilde{r}_{n-1,n} \\ \tilde{r}_{n,n} \end{bmatrix} = \begin{bmatrix} \hat{p}_1 - e \\ \cdots \\ \hat{p}_{n-1} - e \\ 0 \\ 0 \end{bmatrix}, \quad \mathbf{v} = \dot{\mathbf{r}} = \begin{bmatrix} \dot{p}_1 - \dot{e} \\ \cdots \\ \dot{p}_{n-1} - \dot{e} \\ 0 \\ 0 \end{bmatrix}, \quad \mathbf{u} = \begin{bmatrix} u_1 \\ \cdots \\ u_{n-1} \\ u_n \end{bmatrix}$$

$$\mathbf{M} = [M_1, \ldots, M_{n-1}, M_n].$$

So, it is possible to rewrite equations (9)–(10) for $i = \overline{1, n}$, $j = n$ as

$$\dot{\mathbf{r}} = \mathbf{v}, \tag{11}$$

$$\mathbf{v} = \mathbf{Mu}. \tag{12}$$

Dutta et al. [4] also suggests for us to use the matrix W with components

$$w_{ij} = \begin{cases} -b_{ij}, i \neq j; \\ \sum_{i=1, i \neq j} b_{ij}, \end{cases}$$

$b_{ij} = \gamma - a_{ij}$, $\gamma \leq 1$, $W = \gamma C - L$ where C is a constant Laplacian-like matrix having all off-diagonal elements as ones, instead of L to measure the net's connectivity.

Let us denote

$$\mathbf{W}_* = (W \otimes I_m),$$

where $\otimes$ is the Kronecker product of matrices. If A is an $m \times n$ matrix and B is a $p \times q$ matrix, then the Kronecker product $A \otimes B$ is the $mp \times nq$ block matrix:

$$A \otimes B = \begin{bmatrix} a_{11}B & \cdots & a_{1n}B \\ \vdots & \ddots & \vdots \\ a_{m1}B & \cdots & a_{mn}B \end{bmatrix}$$

It is necessary to choose Lyapunov candidate function for (11)–(12) as

$$\mathbf{V} = \mathbf{r}^T \mathbf{W}_* \mathbf{r},$$

and proof the global asymptotical stability of the equilibrium point for the problem (11)–(12) for control vector $\mathbf{u}$ with coordinates

$$u_i = \frac{M_i^{-1}}{\sum_{j \in N_i} w_{ij}} \sum_{j \in N_i} w_{ij} [\ddot{p}_j - \alpha_1 (\hat{\dot{p}}_i - \hat{\dot{p}}_j) - \alpha_2 (\dot{p}_i - \dot{p}_j)],$$

using the backstepping technique. There $\alpha_1 = 1 + k_1 k_2$, $\alpha_2 = k_1 + k_2$ are positive backstepping-related constants.

Later, the aforementioned authors propose a more flexible controller

$$u_i = \frac{M_i^{-1}}{\sum_{j \in N_i} a_{ij}(t)} \sum_{j \in N_i} a_{ij}(t) [\ddot{p}_j - \alpha_1 (\hat{\dot{p}}_i - \hat{\dot{p}}_j)^{\gamma_1} - \alpha_2 (\dot{p}_i - \dot{p}_j)^{\gamma_2}], \qquad (13)$$

where $\gamma_1 > 0$, $\gamma_2 > 0$ are controller parameters.

In the proof, the following restriction is placed on the W matrix $h = 2 \min(k_1, k_2)$.

$$\dot{W} < hW, \qquad (14)$$

It is possibly to proof this theorem if every agent a_i has complete awareness about all others and $N_i = V \setminus \{a_i\}$.

The example of trajectories of four UAV accompanying the target in the centre with a complete communication graph, corresponding to the control input (13) which is depicted in the Fig. 2.

3. Potential Field and Supergradient Based Control, Air-craft Trajectory Predictions

The paper [6] considers a multi-agent system 1, that moves on the plane. The dynamic of i–th agent is described as a single-integrator

$$\dot{p}_i = u_i, \qquad (S1)$$

where $u_i \in \mathbb{R}^2$ is the control input of agent a_i. The strength of the connection decreases smoothly with distance. If the distance between two agents is less than a threshold R_s, they are "strongly" connected. Otherwise the connection is exponentially weakened as the distance increases, until the distance R, where they practically lose the connection.

The elements a_{ij} of the adjacency matrix A for this system are as following:

$$a_{ij} = \begin{cases} 1, \|r_{ij}\| < R_s, \\ e^{\frac{-5(\|r_{ij}\| - R_s)}{R - R_s}}, R_s \leq \|r_{ij}\| \leq R, \\ 0, \|r_{ij}\| > R. \end{cases}$$

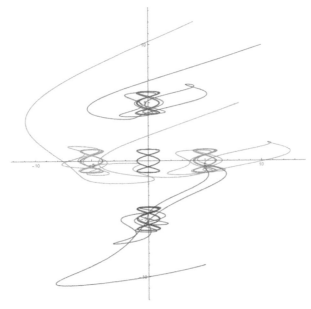

Fig. 2. Trajectories of four UAV accompanying the target
in the centre

The optimization problem is with regards to maximize the connectivity $\lambda_2 (L(\mathbf{p}))$, where $\mathbf{p} = [p_1,..., p_n]^T$, L is the Laplacian matrix.

Definition 1 *Let $f : \mathbb{R}^n \to \mathbb{R}$ be a concave function. The vector g is a super-gradient of f in the point x if for all $y \neq x$ the following inequality holds:*

$$f (y) \leq f (x) + g^T (y - x).$$

To maximize the function f, the updating rule of the supergradient at the step k is: $x^{(k+1)} = x^{(k)} + \alpha^{(k)} g^{(k)}$, where $\alpha^{(k)} > 0$ is the step-size.

The authors, De Gennaro and Jadbabaie [6], obtain a supergradient matrix $G = v_2 v_2^T$ for $\lambda_2 (L)$:

$$\lambda_2(\tilde{L})\lambda_2 (L) + < v_2 v_2^T , (\tilde{L} - L) >,$$

where $\tilde{L} \neq L$, and v_2 is the unit Fiedler vector of L from this definition.
The updating rule for the L matrix is:

$$L^{*(k+1)} = L^{*(k)} + \alpha^{(k)} G^{(k)}.$$

It is known that if the step-size $\alpha^{(k)}$ is the coefficient of a not summable but square summable series, the supergradient method converges to the optimal value. The aforementioned authors had proposed the decentralized supergradient algorithm and the additional potential-based algorithm to minimize errors between the current Laplacian $L(\mathbf{p})$ and the optimum value $L^{*(k)}$. This additional optimization problem can be formulated for the agent a_i as to find p_i to minimize

$$\left\| L_i(\mathbf{p}) - L_i^{*(k)} \right\|_2^2 \tag{15}$$

where $L_i(\mathbf{p})$ is the i–th row of the Laplacian matrix, as a function of the state of the agents, and $L_i^{*(k)}$ is the row of the optimal Laplacian found by agent a_i at the step k of the supergradient. The minimization problem is solved by using potential functions. From each $L_{ij}^{*(k)}$, the authors obtain the desired distance $\delta_{ij}^{(k)}$ between connected agents a_i, a_j:

$$\delta_{ij}^{(k)} = a_{ij}^{-1}(L_{ij}^{*(k)}).$$

For each pair of connected agents a_i, a_j a quadratic potential function $V_{ij}(\| r_{ij} \|)$ is defined as:

$$\begin{cases} (\left\| r_{ij} \right\| - \delta_{ij}^{(k)})^2, \left\| r_{ij} \right\| \leq R, \\ (R - d_{ij}^{(k)})^2, \left\| r_{ij} \right\| > R. \end{cases}$$

The potential function $V_{ij}(\| r_{ij} \|)$ is a positive definite, it is zero when the distance $\| r_{ij} \|$ is equal to the desired value $\delta_{ij}^{(k)}$, and it becomes constant when the distance $\| r_{ij} \|$ is greater than the connection radius R. The control action for agent a_i is defined by solving the optimization problem, equivalent to (15):

$$\min_{p_i} \sum_{j \in N_i} V_{ij}$$

from which the control action on agent a_i is defined as the sum of the negative gradients of the potentials V_{ij} for all $j \in N_i$:

$$u_i = - \sum_{j \in N_i} \nabla_{p_i} V_{ij}.$$

Each agent a_i moves according to this control action, until an equilibrium for the group is reached. At the equilibrium, agent a_i stops and sends the information about its current position $p_i(t)$ to the neighbours, and receives their current positions. With this information, agent a_i updates the row i of the Laplacian matrix $L(\mathbf{p})$. This updated row will be used from agent a_i at the step $k + 1$ of the supergradient.

The potential based control applied by the agents does not ensure the minimization of (15), because it presents local minima if the graph of connections of the group is not a tree.

The work [2] proposed method is used for multi-UAV control without any leader and provides multiple conflicts resolution. The UAV in this article have linear dimensions which characterizes with its ellipsoidal protection zone. The authors use the potential field based control strategy.

The classical approach also exists for collision detection. This method is called geometric, in which aircraft trajectory predictions are based on the linear projections of the current vehicle states [7, 15]. The major disadvantage is that prediction errors are negligible only for short time periods and require high rate of surveillance information to update.

We will use further these prediction methods not for collision detection but for its ability to maintainnet connectivity.

4. Telecommunication Network with Directional Antennae

The agents with directional antennae can organize a link only if their antennae have a special position in relation to each other (see Fig. 3).

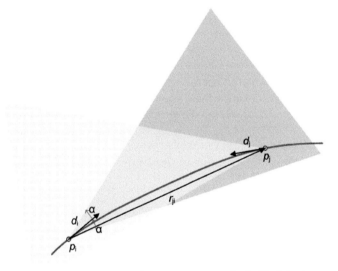

Fig. 3. The pair of agents with coordinates p_i, p_j, α is the half of the beamwidth of the antenna's main lobe.

Let us denote

$$\beta_{ij} = \arccos \frac{(d_i, r_{ji})}{\|d_i\| \|r_{ji}\|}$$

where d_i is the direction of the directional antenna of the i–th agent, $i = \overline{1, n}$.

It seems that we can define the elements of the adjacency matrix $A = (a_{ij})$ of the telecommunication graph as the following

$$a_{ij} = \begin{cases} 1, & \|r_{ij}\| < R \wedge \beta_{ij} < \alpha, \\ 0, & \|r_{ij}\| \geq R \vee \beta_{ij} \geq \alpha. \end{cases}$$

Unfortunately, the adjacency matrix defined in such a way is not symmetric. So we should use

$$a_{ij} = \begin{cases} 1, & \|r_{ij}\| < R \wedge \beta_{ij} < \alpha \wedge \beta_{ji} < \alpha, \\ 0, & \|r_{ij}\| \geq R \vee \beta_{ij} \geq \alpha \vee \beta_{ji} \geq \alpha. \end{cases}$$

Also, we should introduce special distance to define an agent's neighbourhood:

$$N_i = \{j \in V, j \neq i: a_{ij} \neq 0\}.$$

The main difficulty of this network is that it is not possible to maintain connectivity when the number of agents are $n > 2$. Moreover, in certain periods of the network functioning no agent can detect "target", so cooperative formation maintaining previously described is not possible (see Fig. 4).

Finally, if we use the algorithms from the review part of the present work, at the moment of changing the network connectivity, the agent, who has already determined the position of the target and is trying to take the necessary position relative to the target, begins to focus on agents who may be much further from the necessary position. As a result, consensus is never reached.

For example, the agent a_1 in the time moment t_0 (Fig. 4(a)) has the control input

$$u_1 = M_1^{-1}[\ddot{p}_5 - \alpha_1(\hat{p}_1 - \hat{p}_5) - \alpha_2(\dot{p}_1 - \dot{p}_5)]$$

In the time moment t_1 the control should be

$$u_1 = M_1^{-1}[\ddot{p}_2 - \alpha_1(\hat{p}_1 - \hat{p}_2) - \alpha_2(\dot{p}_1 - \dot{p}_2)]$$

but agent a_2 has no information about the target and moves randomly, makes a small circle or waits. So this control leads further from the desired position.

The solution is to compare the distances $s_{ij} = \|\hat{p}_i - \hat{p}_j\|$ and $s_{ik} = \|\hat{p}_i - \hat{p}_k\|$ where a_j is the previously followed agent and a_k is the agent to be followed. If $s_{ij} > s_{ik}$, the agent a_i starts to follow a_k, else a_i constructs the linear prediction C_{t_1} of the a_j trajectory and follows this prediction

$$C_{t_1}(p_j)(t) = p_j(t_1) + v_j(t_1)t,$$

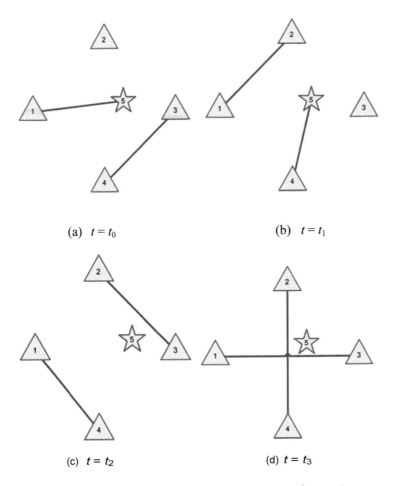

(a) $t = t_0$ (b) $t = t_1$

(c) $t = t_2$ (d) $t = t_3$

Fig. 4. Change in UAVs telecommunication network over time.

$$\hat{p}_j(t) = C_{t_1}(p_j)(t) - R_j$$

until the next moment of connectivity change.

At each moment, where the communication link between agents a_i and a_j is accessible, agents exchange with the newest information about the positions of all agents in the network and synthesize the new controller according to this information.

The aforesaid can be summarized as the following rules for the agent a_i:

1. In the initial moment, the control input is

$$u_i = \frac{M_i^{-1}}{\sum_{j \in N_i} a_{ij}} \sum_{j \in N_i(0)} a_{ij}[\ddot{p}_j - \alpha_1(\hat{p}_i - \hat{p}_j) - \alpha_2(\dot{p}_i - \dot{p}_j)]$$

if N_i is non-empty. Otherwise, $u_i = 0$. Agents sets $n \times n$ array MT (a_i) to $(0, 0, 0)$.

2. When a_i meets a_j in the position p_j at the time mt_{ij}, it sets MT_{ij} $(a_i) = MT_{ji}$ $(a_i) = (p_j, \dot{p}_j, mt_{ij})$. Also, they exchange with MT matrices. If element MT_{kl} (a_j) is newer than the corresponding element MT_{kl} (a_i) then a_i sets $MT_{kl}(a_i) = MT_{kl}(a_j)$.

3. If $a_{ij} = 0$ in time moment t, but $MT_{ij}(a_i) \neq (0, 0, 0)$ and MT_{ij} $(a_i) = (p_j^0, v_j^0, t_{ij})$, then a_i assumes $p_j = p_j^0 + v_j^0 t$ and use it in the controller.

Interestingly enough, an attempt to model such a system of agents leads to systems of differential equations that cannot be solved using the standard solvers, which can be find in MATLAB or Wolfram Mathematica. I plan to do further research on such systems and find effective methods for solving them.

References

1. S.G. Bauer, M.O. Anderson and J.R. Hanneman. Unmanned aerial vehicle (uav) dynamic-tracking directional wireless antennas for low powered applications that require reliable extended range operations in time critical scenarios. Technical Report INL/EXT-05-00883, Idaho National Laboratory, October (2005).

2. V. Chepizhenko and I. Skyrda. Method of the multi-uav formation flight control. *In*: V. Ermolayev, M.C. Suárez-Figueroa, A. Lawrynowicz, R. Palma, V. Yakovyna, H.C. Mayr, M.S. Nikitchenko and A. Spivakovsky (eds), Proceedings of the 14th International Conference on ICT in Education, Research and Industrial Applications. Integration, Harmonization and Knowledge Transfer. Volume I: Main Conference, Kyiv, Ukraine. May 14-17, 2018, volume 2105 of CEUR Workshop Proceedings, 167–178. CEUR-WS.org (2018).

3. J.P. Desai, J. Ostrowski and V. Kumar. Controlling formations of multiple mobile robots. *In*: Proceedings. 1998 IEEE International Conference on Robotics and Automation (Cat. No. 98CH36146), volume 4, 2864–2869, May (1998).

4. R. Dutta, L. Sun, M. Kothari, R. Sharma and D. Pack. A cooperative formation control strategy maintaining connectivity of a multi-agent system. *In*: 2014 IEEE/RSJ International Conference on Intelligent Robots and Systems, 1189–1194, Sep. (2014).

5. R. Dutta, Chunjiang Qian, Liang Sun and D.J. Pack. A generic formation controller and state observer for multiple unmanned systems. CoRR, abs/1709.01321 (2017).

6. M.C. De Gennaro and A. Jadbabaie. Decentralized control of connectivity for multi-agent systems. Proceedings of the 45th IEEE Conference on Decision and Control, 3628–3633 (2006).

7. A. Geser and C. Munoz. A geometric approach to strategic conflict detection and resolution [atc]. *In*: Proceedings. The 21st Digital Avionics Systems Conference, volume 1, 6B1–6B1, Oct (2002).

8. C. Godsil and G. Royle. Algebraic Graph Theory, vol. 207 of Graduate Texts in Mathematics. Springer (2001).

9. H. Kawakami and T. Namerikawa. Cooperative target-capturing strategy for multi-vehicle systems with dynamic network topology. *In*: Proceedings of the 2009 Conference on American Control Conference, ACC'09, IEEE Press, Piscataway, NJ, USA, 635–640 (2009).

10. H.K. Khalil. Nonlinear Systems, 3rd ed. Prentice-Hall. Upper Saddle River, NJ (2002). The book can be consulted by contacting: PH-AID: Wallet, Lionel.

11. P.V. Kokotovic. The joy of feedback: Nonlinear and adaptive. IEEE Control Systems Magazine, 12(3), 7–17, June (1992).

12. M.Kothari, R. Sharma, I. Postlethwaite, R.W. Beard and D. Pack. Cooperative target-capturing with incomplete target information. J. Intell. Robotics Syst. 72(3-4), 373–384, December (2013).

13. A.V. Kuznetsov. Organization of the system of agents with the help of cellular automaton. Control of Large Systems, 70, 136–167 (2017).

14. A. Kuznetsov. Self-organization of the communication network. Journal of Physics: Conference Series, 1203:012093 (April 2019).

15. Jung-Woo Park, Hyon-Dong Oh and Min-Jea Tahk. Uav collision avoidance based on geometric approach. 2008 SICE Annual Conference, 2122–2126 (2008).

16. Wei Ren. Multi-vehicle consensus with a time-varying reference state. Systems & Control Letters, 56, 474–483 (2007).

17. R. Sharma, M. Kothari, C.N. Taylor and I. Postlethwaite. Cooperative target-capturing with inaccurate target information. *In*: Proceedings of the 2010 American Control Conference, 5520–5525, June (2010).

18. M.M. Zavlanos and G.J. Pappas. Controlling connectivity of dynamic graphs. Proceedings of the 44th IEEE Conference on Decision and Control, 6388–6393 (2005).

19. M.M. Zavlanos, S. Grijalva and G.J. Pappas. Graph-theoretic connectivity control of mobile robot networks. Proceedings of the IEEE, 99, 1525–1540 (2011).

10

Induction and Physical Theory Formation As Well As Universal Computation by Machine Learning

Alexander Svozil[1] and Karl Svozil[2]

[1] Theory and Applications of Algorithms, Faculty of Computer Science, University of Vienna Waḧringer Straße 29, A-1090 Vienna, Austria
[2] Institute for Theoretical Physics, Vienna University of Technology, Wiedner Hauptstraße 8-10/136, A-1040 Vienna, Austria

1. Algorithmic Induction

There appears to be at least two approaches towards induction. The first route is by intuition and ingenuity. This route has been successfully pursued by geniuses and gifted individuals. A typical representative of this approach to knowledge is Ramanujan who seemed to have attributed his revelations to a Hindu Goddess [1]. In western thought, this is often more secularly referred to as Platonism. Gödel appeared to have held the opinion that our minds have access to the truth, which can be discovered through personal insights – perhaps even beyond the bounds of universal computability – in particular, the idea that minds are no (Turing) machines [2, p. 216]. As successful as these narratives may have been, they remain anecdotal and cannot be generalized.

When it comes to *ad hoc* revelations of individuals, there may also be psychological issues. These have been described by Freud [3], as pointing to the dangers caused by "temptations to project, what [the analyst] in dull self-perception recognizes as the peculiarities of his own personality, as generally valid theory into science." A similar warning comes from Jaynes' "Mind Projection Fallacy" [4, 5], pointing out that *we are all under an ego-driven temptation to project our private thoughts out onto the real world, by supposing that the creations of one's own imagination are real*

---
*Corresponding author: alexander.svozil@nivie.ac.at, svozil@tuwien.ac.at

properties of Nature, or that one's own ignorance signifies some kind of indecision on the part of Nature."

A second, computational, approach could be conceived in the spirit of Turing [6]. In this line of thought, it is possible to obtain knowledge about a system by mechanical, algorithmic procedures; such as a deterministic agent *"provided with paper, pencil and rubber, and subject to strict discipline [carrying out a set of rules of procedure written down]"* [7, p. 34]. Recently one of the more promising methods to algorithmic induction has been machine learning [8] which will be pursued in this chapter. There is even a form of collective intuition – the so-called *swarm intelligence* [9, 10, 11] – that can solve many logistic problems effectively.

Two *caveats* should be stated upfront. First, that the general induction problem is – in recursion theoretic terms, the rule inference problem – unsolvable with respect and relative to universal computational capacities [12, 13, 14, 15]. So, in certain (even constructible) situations machine learning, like all other algorithmic induction strategies, provably fails. But that does not exclude heuristic methods of induction, such as machine learning applied to physical phenomena. Secondly, theoretical constructions cannot be expected to faithfully represent the "laws underlying" a phenomenology. As Lakatos [16] points out, the progressiveness and degeneracy of research programs are transient; and without a recognizable coherent conceptual convergence. Therefore, the "explanations" and theoretical models generated by machine learning present knowledge and explanations which cannot claim to have any ontologic (only epistemic) relevance – they are means relative to the respective methods employed.

Whereas machine learning has already been applied to very specific problems in high energy [17] and solid-state physics [18], we would like to propose this as an extremely general method of theory formation and induction.

We are applying the obtained machine representations to predict a simulacrum – we are, in particular, interested in universal Turing computation; at least until a finite amount of computational space and time [19].

2. Linear Models of Inertial Motion

In what follows a *linear regression* model (Sect. 5.1.4) [8], will be used. Thereby we shall, if not stated otherwise, explicitly closely follow the notation of Mermin's book on *Quantum Computer Science* [20].

Suppose, for the sake of demonstration, a one-dimensional physical system of a particle in inertial motion. Suppose further that it has been (approximately) measured already at $n \geq 2$ positions $x_1, \ldots, x_n$ at times $t_1, \ldots, t_n$, respectively. The goal is to find a general algorithm which predicts its location at an arbitrary time T.

In what follows the respective positions and times are (not necessarily successively) arranged as n-tuples; that is, as a finite ordered list of elements, and interpreted as $(n \times 1)$-matrices

$$|x\rangle \equiv (x_{i_1}, \ldots, x_{i_n})^T$$
$$|t\rangle \equiv (t_{i_1}, \ldots, t_{i_n})^T \tag{1}$$

where superscript T indicates transposition, and $i_1, \ldots, i_n$ are arbitrary permutations of $1, \ldots, n$. That is, it is not necessary to order the events temporally; actually, they can "run backward" or be randomly arranged [21].

A linear regression *Ansatz* is used to find a linear model for the prediction of some unknown observable, given some of the anecdotal instances of its performance. More formally, let y be an arbitrary observable which depends on n parameters $x_1, \ldots, x_n$ by linear means; that is, by

$$y = \sum_{i=1}^{n} x_i r_i = \langle \mathbf{x} | \mathbf{r} \rangle \tag{2}$$

where $\langle \mathbf{x} | = (|\mathbf{x}\rangle)^T$ is the transpose of the vector $|x\rangle$, the tuple

$$|\mathbf{r}\rangle = (r_1, \ldots, r_n)^T \tag{3}$$

contains the unknown weights of the approximation – the "theory," if you like – and $\langle \mathbf{a} | \mathbf{b} \rangle = \sum_i a_i b_i$ stands for the Euclidean scalar product of the tuples interpreted as (dual) vectors in n-dimensional (dual) vector space $\mathbb{R}^n$.

Given are m known instances of (2); that is, suppose m pairs are $(z_{j_i}, |\mathbf{x}_j\rangle)$ known. These data can be bundled into an m-tuple

$$|\mathbf{z}\rangle \equiv (z_{j_1}, \ldots, z_{j_m})^T \tag{4}$$

and an $(m \times n)$-matrix

$$\mathbf{X} \equiv \begin{pmatrix} x_{j_1 i_1} & \cdots & x_{j_1 i_n} \\ \vdots & \vdots & \vdots \\ x_{j_m i_1} & \cdots & x_{j_m i_n} \end{pmatrix} \tag{5}$$

where $j_1, \ldots, j_m$ are arbitrary permutations of $1, \ldots, m$, and the matrix rows are just the vectors $|\mathbf{x}_{j_k}\rangle \equiv (x_{j_k i_1}, \cdots, x_{j_k i_n})^T$.

The task is to compute a "good" estimate of $|\mathbf{r}\rangle$; that is, an estimate of $|\mathbf{r}\rangle$ which allows for an "optimal" computation of the prediction y.

Suppose that a good way to measure the performance of the prediction from some particular definite but unknown $|r\rangle$ with respect to the m given data $(z_j, |x_j\rangle)$ is by the mean squared error (MSE)

$$
\begin{aligned}
\frac{1}{m}\||y\rangle - |z\rangle\|^2 &= \frac{1}{m}\|X|r\rangle - |z\rangle\|^2 \\
&= \frac{1}{m}(X|r\rangle - |z\rangle)^T (X|r\rangle - |z\rangle) \\
&= \frac{1}{m}(\langle r|X^T - \langle z|)(X|r\rangle - |z\rangle) \\
&= \frac{1}{m}(\langle r|X^T X|r\rangle - \langle z|X|r\rangle - \langle r|X^T|z\rangle + \langle z|z\rangle) \\
&= \frac{1}{m}\left[\langle r|X^T X|r\rangle - \langle z|(\langle r|X^T)^T - \langle r|X^T|z\rangle + \langle z|z\rangle\right] \\
&= \frac{1}{m}\left\{\langle rX^T X|r\rangle - [(\langle r|X^T)^T]^T|z\rangle - \langle r|X^T|z\rangle + \langle z|z\rangle\right\} \\
&= \frac{1}{m}(\langle r|X^T X|r\rangle - 2\langle r|X^T|z\rangle + \langle z|z\rangle) \ .
\end{aligned}
\tag{6}
$$

In order to minimize the mean squared error (6) with respect to variations of $|r\rangle$ one obtains a condition for "the linear theory" $|y\rangle$ by setting its derivatives (its gradient) to zero; that is

$$
\partial_{|r\rangle}\mathrm{MSE} = \mathbf{0}.
\tag{7}
$$

A lengthy but straightforward computation yields

$$
\frac{\partial}{\partial r_i}(r_j X^T_{jk} X_{kl} r_l - 2r_j X^T_{jk} z_k + z_j z_j)
$$

$$
\begin{aligned}
&= \delta_{ij} X^T_{jk} X_{kl} r_l + r_j X^T_{jk} X_{kl}\delta_{il} - 2\delta_{ij} X^T_{jk} z_k \\
&= X^T_{ik} X_{kl} r_l + r_j X^T_{jk} X_{ki} - 2X^T_{ik} z_k \\
&= X^T_{ik} X_{kl} r_l + X^T_{ik} X_{kj} r_j - 2X^T_{ik} z_k \\
&= 2X^T_{ik} X_{kj} r_j - 2X^T_{ik} z_k \\
&= 2(X^T X|r\rangle - X^T|z\rangle) = 0
\end{aligned}
\tag{8}
$$

and finally, upon multiplication with $(X^T X)^{-1}$ from the left,

$$
|r\rangle = (X^T X)^{-1} X^T|z\rangle.
\tag{9}
$$

A short plausibility check for $n = m = 1$ yields the linear dependency $|z\rangle = X|r\rangle$.

Coming back to the one-dimensional physical system of a particle in inertial motion, we could characterize inertial motion in machine learning and linear regression terms by the requirement that the Ansatz (2) is "good" in the sense that predictions can be made to a "sufficient degree" (a term which is arbitrary, subjective and thus conventional); that is, within a pre-defined error.

If the particle does not pass through the origin, it might be necessary to augment Eq. (2) with an affine term d, which can be absorbed into $|x'\rangle = (x_1', ..., x_{n+1}') = (x_1, ..., x_n, 1)^T$ and $|r'\rangle = (r_1', ..., r_{n+1}') = (r_1, ..., r_n, d)^T$ such that

$$y = \sum_{i=1}^{n} x_i r_i = \langle x | r \rangle + d = \sum_{i=1}^{n+1} x_i' r_i' \tag{10}$$

3. Nonlinear Models of Noninertial Motion

The linear Ansatz (2) fails for noninertial motion. One possible way to cope with nonlinearities would be to introduce extra dimensions corresponding to nonlinear terms, such as x^l for $2 \leq l \leq d < \infty$; with the consequence that the dimensionality of the parameter space increases.

To cope with nonlinear phenomena, *deep forward networks* have been used in machine learning [8, Chapt. 6]. This strategy of deep learning invokes intermediate *hidden theoretical layers* of description which communicate with each other. For the sake of an example, suppose there are two functions g and h, connected in a chain by functional substitution, such that $f(x) = h(g(x))$. The length of the chain is identified with the *depth* of the model – in this case two. g is the first layer of the model. The final layer – in this case h – is called the *output layer*.

Unfortunately, the linear regression Ansatz (10) for g and h would effectively be linear again. Therefore, to model a nonlinear phenomenology, a nonlinear Ansatz for at least one layer has to be implemented. Such networks are capable of approximating any Borel measurable function (and its derivative, even if it is a generalized function) from one finite-dimensional space to another [22, 23, 24].

4. Simulation of Universal Turing Machines by Deep Forward Networks

It could be suspected that, even though for all practical purposes [25] the methods and techniques discussed so far are "good," the task of finding "better and better" approximations or even total correspondence might, at least in some cases, turn out to be "difficult" if not outrightly impossible. Because suppose, for the sake of a *reductio ad absurdum* – more precisely, a reduction to the rule inference and halting problems

– that such learning of the exact behavior would be computable. Such a suspicion might be tempting, considering the approximate "solution" of the general rule inference problem [12, 13, 14, 15] suggested earlier in the limit of infinite precision. Alas, any such unbounded computation, as long as it needs to be finite, would run into the problem that no computable rate of convergence can be given – very much like the Busy Beaver function [26] or Chaitin's halting probability Ω [27, 28]. Formally, any such claim can be reduced to the rule inference and the halting problem of universal computers. The situation is not dissimilar to series solutions of the n-body problem [29], which may converge "very slowly" (indeed, intractably slow in numerical work [30]); but if the system encodes a universal computer, they cannot converge in general due to the reduction to the halting problem [31].

However, one could (courageously) "invert" or transform these objections and deficiencies into virtues [32, 33] and argue that, for all practical purposes and in many relevant instantiations, machine learning, and, in particular, deep forward networks, may turn out to be effective in the simulation of universal computers such as a universal Turing machine. Thereby the criterion is not to obtain an exact correspondence; rather the result of the simulation should be "good enough" to justify its use.

For the sake of an example, think of the typical physical estimate of a quantity in terms of *orders of magnitude*: very often, the applicability of a suggestion or technique does not depend on the exact observable value it generates but rather on the order of its impact. This is not dissimilar to certain quantum advantages: for instance, the Deutsch algorithm identifies the parity or (non)constancy of a binary function of a single bit without identifying the actual function [20]; thereby rendering a partitioning of the set of all such individual functions.

How could one imagine training a deep forward network to simulate a universal computer? Fairly simple: let it access the input-output behavior of actual "exact" devices with von Neuman architecture. After "lots of" training, the network should be able to emulate the performance of this architecture within "reasonable precision." That is, it won't be able to give the exact value of, say, an algebraic operation like $n + m$ for "large" (for physical realizability) numbers n and m. But it might be able to output some estimates which are "close to" (for the applicability) the exact result.

One could also say that, in this scheme, the deep forward network acts as an oracle with respect to the universal computer. And although the training of it may take some time because of the sheer training volume, as well as the conceivable computational complexity of the individual input-output functions involved, eventually, the trained network is not bound by these restrictions and can reach an (approximate) result quite fast; such that the computation time it takes for any simulation is uniformly bounded from above. In particular, the halting problem can be said to be "for all practical purposes (FAPP) solvable" by such oracles; but, of course, no guarantee of validity or total precision can be given. Inadvertently one cannot exclude instances in which the deep forward network predicts halting whereas the universal computer it simulates does not halt, and vice versa.

One may also ask: where exactly is the physical resource rendering the computational capacity of such "universal" deep forward networks located? There has to be some formal symbolic encoding in terms of physical components or entities making the simulation feasible and effective. One rather straightforward way to answer this is in terms of the *connections* or *correlations* among the nodes involved: if they are modeled as continuous formal entities such as real or complex numbers then the capacity of even a finite such configuration to store information is unbounded.

In more pragmatic, practical terms the non-exact but effective simulation of general (universal) computations may also present a way to circumvent the stall in Moore's law which can be observed already for a couple of years. Currently, because of physical restrictions on circuit and switch designs, most improvements in performance are due to parallelization rather than miniaturization. Eventually, the switching time of electronic devices is restricted by fundamental limits from below on resistance; in particular the von Klitzing constant R_k.

5. Discussion

One objection for applying machine learning algorithms to physical theory creation or simulation of universal computers might be that the resulting representations lack any sort of "meaning;" that is, these representations amount to pure syntax devoid of any conceptual semantics. But if conceptual semantics is omitted, there can be no true "understanding" of the "physics behind" the phenomena, or the computational processes yielding those estimates.

One may counter this criticism by noticing that, first, underlying such objections is the premise that something can actually be discovered or revealed. This realistic ontology is by far from non-trivial and is heavily debated [34, 35, 36]. If, for example, the phenomena emerges from primordial chaos, such as in Greek mythology and cosmology, χαος, then any "meaning" one might present and "discover" ultimately remains a (pragmatic) narrative or a mathematical abstraction such as Ramsey theory at best: for any data, there cannot be no correlations – regardless of the origin or type of empirical data, there always has to be some, maybe spurious [37, 38], regularity or coincidences or properties.

How can it be excluded that the laws of physics mean nothing but yet undiscovered consequences of Ramsey theory?

Second, as has already been pointed out, historic evidence seems to suggest that successive physical conceptual models (say, of gravity) are not continuously evolving; but that they are disruptive and dissimilar [39, 16]: they lack conceptual convergence. One may even go so far as to suggest that, in any case, theories are (more or less [40]) successful belief systems; very much like Greek mythology [41].

Third, also the present perception of the quantum mechanical formalism includes, among other inclinations, the position that no interpretation is necessary [42]; that indeed, interpretation is even dangerous and detrimental for the researcher [43, p. 129]; or that, at the very least, there are no issues with respect to interpretation [44].

Nevertheless, it might be quite amusing to study toy universes capable of universal computation, such as Conway's *game of life*, via intrinsic, embedded, machine learning algorithms. It cannot be excluded that these kinds of algorithmic agents "come up" with the "right rules;" that is those rules which define the toy mini-universe. It can be expected that if a machine learning algorithm performs excellently on particular problems then it necessarily has a degraded performance on the set of all remaining problems [45, 46].

In any case, the manner in which physical theories are created and invented by human individuals is not dissimilar from machine learning. And machine learning might become of great practical utility for the simulation of (universal) computations.

Acknowledgments

This work was supported in part by the European Union, Research Executive Agency (REA), Marie Curie FP7-PEOPLE-2010-IRSES-269151-RANPHYS grant. Responsibility for the information and views expressed in this article lies entirely with the authors. The authors declare no conflict of interest.

References

1. R. Kanigel. The Man Who Knew Infinity: A Life of the Genius Ramanujan, 5th ed. Washington Square Press (1991).
2. G. Kreisel. Kurt Gödel. 28 April 1906-14 January 1978. Biographical memoirs of Fellows of the Royal Society 26, 148–224, corrections Ibid. 27, 697; ibid. 28, 718 (1980).
3. S. Freud. Ratschläge für den Arzt bei der psychoanalytischen Behandlung. *In*: Gesammelte Werke. Chronologisch geordnet. Achter Band. Werke aus

den Jahren 1909–1913, edited by Anna Freud, E. Bibring, W. Hoffer, E. Kris and O. Isakower, Fischer, Frankfurt am Main, 1912, 1999, pp. 376–387 (1999).

4. E.T. Jaynes. Clearing up mysteries – the original goal. *In*: Maximum-Entropy and Bayesian Methods: Proceedings of the 8th Maximum Entropy Workshop, held on August 1-5, 1988, in St. John's College, Cambridge, England, edited by John Skilling, Kluwer, Dordrecht, pp. 1–28 (1989).

5. E.T. Jaynes. Probability in quantum theory. *In*: Complexity, Entropy, and the Physics of Information: Proceedings of the 1988 Workshop on Complexity, Entropy, and the Physics of Information, held May-June, 1989, in Santa Fe, New Mexico, edited by Wojciech Hubert Zurek Addison-Wesley, Reading, MA, pp. 381–404 (1990).

6. A.M. Turing. Intelligent machinery, a heretical theory. Philosophia Mathematica 4, 256–260 (1996).

7. A.M. Turing. Intelligent machinery. *In*: Cybernetics. Key Papers, edited by C.R. Evans and A.D.J. Robertson. Butterworths, London, pp. 27–52 (1968).

8. I. Goodfellow, Y. Bengio and A. Courville. Deep Learning (2016).

9. E. Bonabeau, M. Dorigo and G. Theraulaz. Swarm Intelligence: From Natural to Artificial Systems. Santa Fe Institute Studies on the Sciences of Complexity. Oxford University Press, New York, NY (1999).

10. J. Kennedy, R.C. Eberhart and Y. Shi. Swarm Intelligence. The Morgan Kaufmann Series in Evolutionary Computation. Morgan Kaufmann, Academic Press. Elsevier, San Francisco, San Diego, CA (2001).

11. I. Zelinka and G. Chen. Evolutionary Algorithms, Swarm Dynamics and Complex Networks. Methodology, Perspectives and Implementation, Emergence, Complexity and Computation. Vol. 26. Springer-Verlag, Berlin, Heidelberg (2018).

12. M.E. Gold. Language identification in the limit. Information and Control. 10, 447–474 (1967).

13. L. Blum and M. Blum. Toward a mathematical theory of inductive inference. Information and Control 28, 125–155 (1975).

14. Dana Angluin and Carl H. Smith. Inductive inference: Theory and methods. ACM Computing Surveys 15, 237–269 (1983).

15. L.d.M. Adleman and M. Blum. Inductive inference and unsolvability. The Journal of Symbolic Logic 56, 891–900 (1991).

16. I. Lakatos. Philosophical Papers. 1. The Methodology of Scientific Research Programmes. Cambridge University Press, Cambridge (1978).

17. Kaggle Team. The HiggsML challenge: When high energy physics meets machine learning. May-September 2014, accessed August 31, 2016.

18. L.-F. Arsenault, O. Anatole von Lilienfeld and A.J. Millis. Machine learning for many-body physics: Efficient solution of dynamical mean-field theory. arXiv: 1506.08858 (2015).

19. R.O. Gandy. Limitations to mathematical knowledge. *In*: Logic colloquium '80, edited by D. van Dalen, D. Lascar and J. Smiley. North Holland, Amsterdam, papers intended for the European Summer Meeting of the Association for Symbolic Logic, pp. 129–146 (1982).

20. D.N. Mermin. Quantum Computer Science. Cambridge University Press, Cambridge (2007).
21. G. Egan. Permutation City (1994). Accessed January 4, 2017.
22. K. Hornik, M. Stinchcombe and H. White. Multilayer feed-forward networks are universal approximators. Neural Networks 2, 359–366 (1989).
23. K. Hornik, M. Stinchcombe and H. White. Universal approximation of an unknown function and its derivatives using multilayer feedforward networks. Neural Networks 3, 551–560 (1990).
24. K. Hornik. Approximation capabilities of multilayer feed forward networks. Neural Networks 4, 251–257 (1991).
25. J. S. Bell. Against 'measurement'. Physics World 3, 33–41 (1990).
26. G.J. Chaitin. Computing the busy beaver function. *In*: Thomas M. Cover and B. Gopinath (eds), Open Problems in Communication and Computation. p. 108, Springer, New York (1987).
27. C.S. Calude, M.J. Dinneen and Chi-Kou Shu. Computing a glimpse of randomness. Experimental Mathematics 11, 361–370 (2002), arXiv:nlin/0112022.
28. C.S. Calude and M.J. Dinneen. Exact approximations of omega numbers. International Journal of Bifurcation and Chaos 17, 1937–1954 (2007), CDMTCS report series 293.
29. Qui Dong Wang. The global solution of the n-body problem. Celestial Mechanics 50, 73–88 (1991).
30. F. Diacu. The solution of the n-body problem. The Mathematical Intelligencer 18, 66–70 (1996).
31. K. Svozil. Omega and the time evolution of the n-body problem. *In*: Cristian S. Calude (ed.), Randomness and Complexity, from Leibniz to Chaitin. pp. 231–236. World Scientific, Singapore (2007) arXiv:physics/0703031.
32. F. Nietzsche. Jenseits von Gut und Böse (Beyond Good and Evil) (1886, 2009) digital critical edition of the complete works and letters, based on the critical text by G. Colli and M. Montinari, edited by Paolo D'Iorio. Berlin/New York, de Gruyter (1967).
33. F. Nietzsche. Zur Genealogie der Moral (On the Genealogy of Morality) (1887, 2009) digital critical edition of the complete works and letters, based on the critical text by G. Colli and M. Montinari, edited by Paolo D'Iorio. Berlin/New York, de Gruyter (1967).
34. G. Berkeley. A Treatise Concerning the Principles of Human Knowledge. (1710).
35. W.T. Stace. The refutation of realism. *In*: Herbert Feigl and Wilfrid Sellars (eds), Readings in Philosophical Analysis. pp. 364–372. Appleton-Century-Crofts, New York, 1949, previously published in Mind 53, 349–353 (1934).
36. T. Goldschmidt and K.L. Pearce. Idealism: New Essays in Meta-physics. Oxford University Press, Oxford, UK (2017, 2018).
37. C.S. Calude and G. Longo. The deluge of spurious correlations in big data. Foundations of Science 1–18, CDMTCS-488 (2016).

38. C.S. Calude and K. Svozil. Spurious, emergent laws in number worlds. Philosophies 4, 17 (2019), arXiv:1812.04416.
39. T.S. Kuhn. The Structure of Scientific Revolutions, 2nd ed. Princeton University Press, Princeton, NJ (1970).
40. P.K. Feyerabend. Against Method. New Left Books, London (1974).
41. P. Veyne. Did the Greeks Believe in Their Myths? An Essay on the Constitutive Imagination. University of Chicago Press, Chicago (1988).
42. C.A. Fuchs and A. Peres. Quantum theory needs no 'interpretation'. Physics Today 53, 70–71 (2000), further discussions of and reactions to the article can be found in the September issue of Physics Today, 53, 11-14 (2000).
43. R.P. Feynman. The Character of Physical Law. MIT Press, Cambridge, MA (1965).
44. B.-G. Englert. On quantum theory. The European Physical Journal D 67, 1–16 (2013), arXiv:1308.5290.
45. D.H. Wolpert. The lack of a priori distinctions between learning algorithms. Neural Computation 8, 1341–1390 (1996).
46. D.H. Wolpert and W.G. Macready. No free lunch theorems for optimization. IEEE Transactions on Evolutionary Computation 1, 67–82 (1997).

Index

About the Editor

Education

He received his M.S. and Ph.D. in Philosophy from the Belarusian State University (Minsk, Belarus) in 1998 and 2003, respectively.

Experience

He worked as a Associate Professor at the Belarusian State University, Belarus. Now he works as the Head of the Department of Cognitive Science and Mathematical Modelling at the University of Information Technology and Management in Rzeszow, Poland. He participated in the project *Physarum Chip: Growing Computers from Slime Mould* supported by the Seventh Framework Programme (FP7-ICT-2011-8). The main objectives were to design and fabricate a distributed biomorphic computing device built and operated by slime mould *Physarum polycephalum*. A Physarum chip is a network of processing elements made of the slime mould's protoplasmic tubes coated with conductive substances; the network is populated by living slime mould. A living network of protoplasmic tubes acts as an active non-linear transducer of information, while templates of tubes coated with conductor act as fast information channels.

Research interests

His research focuses on logic and the philosophy of science with an emphasis on non-well-founded phenomena: self-references and circularity. He has contributed primarily to research areas such as reasoning under uncertainty, probability reasoning, non-Archimedean mathematics, as well as their applications to cognitive science. He is also engaged in unconventional computing, decision theory, logical modelling of economics, and the history of logic.

Some Publications

He has authored many books such as *Behaviourism in Studying Swarms: Logical Models of Sensing and Motoring*, Book Series: Emergence, Complexity and Computation. Springer International Publishing (2019); *Talmudic Logic*. Book series: Studies in Talmudic Logic. London: College Publications (2012), and many papers.